EXPERIMENTAL BIOLOGY MANUAL

by
G. D. BROWN, B.Sc., M.I.Biol.
and
J. CREEDY, B.Sc., M.I.Biol.

HEINEMANN EDUCATIONAL BOOKS LTD
LONDON

Heinemann Educational Books Ltd

LONDON EDINBURGH MELBOURNE TORONTO
AUCKLAND SINGAPORE HONG KONG
JOHANNESBURG IBADAN NAIROBI

SBN 435 60188 1

Subj Prd ✓

Published by Heinemann Educational Books Ltd
48 Charles Street, London W1X 8AH
Printed in Great Britain by
Butler & Tanner Ltd
Frome and London

EXPERIMENTAL BIOLOGY MANUAL

Preface

Of recent years there has been an increased interest in the teaching of genetics, cytology, microbiology, biochemistry and physiology. This has gained momentum from the emphasis which has been placed on the integration of the biological subjects for some years now. In this book we are attempting to place between two covers a collection of experiments on these topics. Our aims in doing this are to provide the basis for a pre-university course for future students of all the biological sciences, and for a course which is representative of the present state of knowledge in the subject for pupils who will not continue their study of the subject beyond school level.

The selection of experiments included in this book results from a firm belief that if a pupil undertakes an exercise, he should be permitted to complete it and get meaningful results. To aim at less is to reduce the exercise to a mere practising of techniques. With this in mind each pupil should be permitted to repeat some of the experiments as it is assumed that each pupil will have difficulty with some experiments. Similarly because of differences of ability, interest and luck, each pupil works at a different pace. Thus pupils must work individually as often as possible. This will tax the financial resources of many schools, but it can often be overcome by arranging for pupils to rotate around a series of experiments. We believe that one experiment successfully completed by the individual is worth several shared experiments or demonstrations.

As great emphasis is placed on a pupil getting results which he can interpret, it is essential that every pupil should be able to complete most experiments in a normal school double period. Some chromatography experiments may need to be left to run overnight, while genetics and microbiology experiments may take some days or even weeks. It is sometimes useful if two consecutive double periods can be timetabled for one experiment. In Biology there are many techniques to be mastered before useful results can be obtained. The first double period may be taken up merely mastering the technique.

One unavoidable consequence of the rota arrangement is that it cannot be assumed that a pupil will have received the appropriate theoretical introduction before undertaking an experiment. We have therefore preceded many experiments with a brief theoretical account. A bibliography is provided to supplement this account.

We must emphasise that this is a collection of experiments from which teachers can select appropriate ones for their own courses. These experiments are intended to be supported by suitable readings and discussions.

The actual selection of experiments has been determined by several factors. We are assuming that pupils today enter the sixth form with some experience of experimental work. Many of the more mundane experiments will therefore have been performed as part of pre-sixth form courses, or at least pupils should have performed sufficient experiments to be able to accept other accounts as being accurate. Many other topics, such as histology and dissection, which must still be included in modern courses, are so well dealt with elsewhere that there is no justification for increasing the cost of this book by their inclusion. Nevertheless there are some topics we would like to have included. These have been omitted, either because there seems to be no suitable practical work or because the authors have no first-hand experience of the material in the teaching situation.

We have varied the style of the writing throughout the book. We feel that all the experiments could have been written in such a way that the pupil has to find the answer for himself, but it is useful for the average pupil to have some confidence-producing experiments, ones in which he has some idea of the results before he starts, and from which he can gauge his ability or lack of ability.

As many techniques are used several times throughout the book, we have described each in general terms immediately preceding its first use.

Some parts of the book may appear to be the practising of techniques for their own sake. These are included to form the tools for project work.

Acknowledgements

We wish to thank various people for permission to reproduce and adapt parts of their work for this book.
P. 46, Dade and Gunnell: *Classwork with Fungi* (Commonwealth Mycological Institute). P. 101, 106, 124, 127, 129, Zwarenstein and Van der Schyff: *Practical Biochemistry* (Livingstone). P. 147, W. O. James: *An Introduction to Plant Physiology* (Clarendon Press, Oxford). P. 250, Strafford: *Plant Physiology* (HEB). P. 295–7, Baron: *Organization in Plants* (Arnold), Evans Electroselenium Ltd: *EEL Colorimeter Manual*, Appendix 10.

Contents

CONTENTS

CONTENTS

MICROSCOPY

The microscope

The microscope and its components

All too frequently one sees a microscope set up so badly that it has no chance of providing the quality of image for which it has been designed. For this reason a brief account of the function of the optical parts of the microscope is included in this book in the hope that better understanding on the part of the student may result in the elimination of some of the most common errors.

The light source

Many students will use a bench lamp of some type in conjunction with a mirror at the base of the microscope. When the microscope is provided with a plano-concave mirror, the plane mirror should be used. The lamp should be placed 20–25 cm from the microscope and positioned or masked so that all light falls below the stage. One of the most frequent faults in setting up a microscope is to allow light to enter the slide from the side and above. Considerable loss of image quality will occur unless the slide is viewed by transmitted light which originates below the stage. The concave mirror should never be used in conjunction with a condenser.

If the microscope is provided with simple built-in illumination or with Köhler illumination, check that it is set up and correctly centred as per makers' instructions.

The objective

The quality of the objective and the care which it receives are the most important factors influencing the performance of the microscope. The shorter the focal length of the objective, the greater its magnifying power will be. Most student microscopes are fitted with 16 mm and 4 mm objectives which give magnifications of $\times 10$ and $\times 40$. Sometimes an oil immersion lens with a focal length of 2 mm and a magnification of $\times 100$ is provided. The focal length of the lens is engraved on the barrel of the objective and on modern lenses it is always given in millimetres. Another figure should also be engraved on the objective barrel, this is the numerical aperture of the lens or N.A. The lens barrel may be engraved 16 mm N.A. 0·28 or $\times 10/0·28$ N.A.

The numerical aperture of the lens describes a very important optical property of the objective. The higher the N.A. for any given focal length, the greater the ability of the lens to resolve detail. As the N.A. increases the amount of light passing through the lens and the resolving power of the lens increase in direct proportion, but the depth of field varies inversely. Lenses of very high N.A. are seldom found on school equipment as they are costly to manufacture. Although a lens may have a high N.A., it cannot give its best resolution unless it is used in conjunction with a substage condenser of matching N.A. The N.A. of a condenser cannot exceed 1·0 unless it is designed to work through oil. Although Abbé condensers are available with numerical apertures greater than 1·0, the basic design allows so many aberrations that the extra light transmitted only produces glare which reduces the performance of the objective.

The working distance of the objective is very much less than the focal length of the lens, e.g. a lens of 2 mm focal length may have a working distance of only 0·12 mm and great care is needed to ensure that it is not racked down on to the slide. On modern microscopes there is an increasing tendency to provide 4 mm and 2 mm objectives with spring loaded mounts which greatly reduce the risk of damage.

Objectives with a focal length of 2 mm or less are used in conjunction with an oil of the same refractive index as glass. By using oil immersion lenses for high magnifications, much better resolution can be obtained as the bending of light rays at two glass/air interfaces is eliminated. It is also logical in oil immersion work to use oil between the slide and the condenser.

Eyepieces

The eyepiece magnifies the image produced by the objective. The total magnifying power of a microscope is the product of the magnification of the objective multiplied by the power of the eyepiece, e.g. $\times 40$ objective used with a $\times 10$ eyepiece gives a total magnification of $\times 400$. A number of specially computed eyepieces are available but are beyond the scope of these notes.

Condensers

In 99 cases out of a 100 when a student fails to obtain good resolution with his microscope, the adjustment of the condenser is at fault. Its function should be thoroughly understood by all microscope users. The use of a condenser is two-fold: (*a*) to illuminate evenly the whole of the field in which the object lies; and (*b*) to provide a cone of light of the correct size and character for a given objective.

The Abbé condenser (or Abbé illuminator) is the type most commonly used in schools. It is essential that it should be fitted with an iris to adjust the cone of light to suit each objective, otherwise the glare which it produces may cancel its beneficial effect. Every school should possess at least one microscope fitted with an achromatic or aplanatic condenser for critical work. Many manufacturers now produce a condenser of this type with a 'flip-top' lens for high-power work, which also gives a wide field for low-power work at the flick of a lever. At the base of the condenser there is often a filter carrier which may be swung in or out of use.

4

Filters

The amount of light passing in to the microscope should be controlled by using a rheostat or by placing neutral density filters in the filter carrier. When good resolution is needed it is essential to focus the substage correctly and to adjust the iris diaphragm to the correct N.A. When the microscope has been correctly set up it obviously cannot have its illumination adjusted by use of the iris or by moving the condenser without some effect on its resolution.

Although not strictly a filter, an opal or ground glass plate may be used to diffuse a small light source, or to give even illumination for an objective with a very low magnification and wide field. Such diffusing screens also act as neutral density filters.

Resolution can be improved to some extent by the use of a blue filter.

Infra red absorption filters (also known as heat filters), may be used to prevent overheating of the specimen.

Coloured filters are used to control contrast and are of particular interest in microphotography. Stains in the specimen absorb some of the light; if a filter of the colour absorbed is used, contrast will be increased. The selection of a filter of the complementary colour will lead to a reduction of contrast.

Filters are also used in fluorescence microscopy but as there is little application in school work at the moment, and as many advances are being made in this field at the present time, anyone interested should seek advice from a microscope firm or consult the specialist literature.

Polarising filters may be used in biology. A filter of polaroid material is placed below the slide and a second filter of polaroid material, called the analyser, is placed in a rotating cap above the eyepiece. Polarised light may be used in the examination of starch grains which appear bright with a dark Maltese Cross, crystalline material in plant and animal cells, cuticles of plants, hair, wool and other natural fibres, decalcified teeth and bone, and muscle fibres.

Setting up the microscope

The biologist should acquire a routine procedure for setting up his microscope. The routine should always be used to ensure that the maximum performance is obtained from the instrument.

1. Always try to work in a position where little direct light falls on the instrument. The worst position is one facing a large window as direct light falling on the slide will seriously affect both contrast and resolution.
2. Place the bench lamp 20–25 cm from the microscope.
3. Open the condenser iris fully and raise the condenser until it is level with the stage.
4. Place a low powered objective in position and adjust the plane mirror until the field is evenly illuminated.
5. Focus the objective on the specimen.

5

6. Focus the condenser. If the lamp is fitted with an iris it should be almost closed and the condenser focused until the iris edge is in sharp focus. Open the lamp iris until its aperture does not impinge on the field of view. If the lamp is not fitted with an iris, place a pencil point on the bulb and rack the condenser until the pencil is sharply focused.
7. The condenser iris is now adjusted to match the N.A. of the objective. The simplest method of making this adjustment is to remove the eyepiece and close the iris until it appears to black out about a quarter of the area of the objective. Replace the eyepiece.
8. If the objective is changed to a higher power, the condenser iris must be adjusted so that its N.A. matches that of the objective.
9. Oil immersion lenses are used for the highest magnification. The microscope tube is racked up and the 2 mm lens is swung into place. Place a drop of oil on the slide and focus downwards with the coarse adjustment until the lens makes contact with the oil. Continue to focus down with the coarse adjustment until the colour or blurred outline of the specimen just appears. Complete the focusing with the fine adjustment.

If the 2 mm objective is spring loaded, it is permissible to rack it down with the coarse focus control until the lens just touches the slide, then focus up with the fine adjustment.

Dark-ground microscopy

In normal microscopy the condenser is used to fill the objective with a solid cone of light and the specimen is examined by means of the light transmitted through it. In dark-ground microscopy the condenser is used to illuminate the slide with very oblique rays of light so that none enters the objective. The field, therefore, appears black and any object on the slide is seen by virtue of the fact that it scatters some of the oblique rays passing through the slide. The light which is scattered by the object passes into the objective and the object appears brightly illuminated against a dark background.

For low-power work with the 16 mm ($\times$10) objective, an Abbé condenser, fitted with a patch stop, may be used. The patch stop is a disc of the correct size to prevent the central rays in the cone of light coming from the condenser from reaching the objective. It may be made as follows:

1. Set up the microscope in the normal way.
2. With the eyepiece removed, adjust the condenser iris until its edge just impinges on the back lens of the objective. A patch

of the same diameter as this aperture is cut from a piece of aluminium foil or the black, light-proof paper used to wrap photographic paper. (Use a sharp cork borer with the paper or foil supported on a piece of firm cork.)

3. Place the patch stop, accurately, in the centre of the filter carrier below the condenser.

For high-power work with the 4 mm and 2 mm lens, a specially designed condenser must be used and an objective with a built-in iris is necessary if the N.A. of the objective is higher than 0·95. It is essential for high power, dark-ground work to place immersion oil between the condenser and the slide. As very little of the light used in the system will ultimately enter the objective, it is necessary to use a powerful source such as a Köhler illuminator. **The thickness of the slide is critical. Modern dark-ground condensers are designed for use with slides 1·2 mm thick.** Slides which differ from this thickness may give concentric, fringe images.

Dark-ground illumination is excellent for the examination of protozoa, rotifers, small crustaceans, motile algae and diatoms.

Setting up the microscope for low power dark-ground illumination

1. Place the bench lamp 20 cm from the microscope, making sure that all its light falls below the stage.
2. Focus the 16 mm objective on the slide.
3. Focus the condenser carefully, leave the iris open.
4. Swing the filter holder with its carefully centred patch stop into position.

Setting up the microscope for high power dark-ground illumination

1. Use Köhler or other high intensity illumination, a special dark-ground condenser in a centering mount and high power objectives with a N.A. lower than 0·95 or objectives fitted with a built-in iris.
2. Open condenser iris fully. Adjust the light source, if necessary, so that the whole of the condenser top appears to be evenly illuminated when viewed from the side.
3. Check that the condenser is correctly centered in its mount.
4. Make sure that the condenser is below the level of the stage and apply a large drop of immersion oil to the centre of the condenser lens.
5. Place the slide in position.
6. Raise the condenser until the oil makes contact with the slide and spreads to the edge of the condenser lens.
7. Focus the 16 mm objective. You should see a bright ring with a dark centre. Make a final check that the condenser is correctly centered with the centre of the dark spot in the optical axis of the microscope.

8. Focus the condenser up or down very slowly until the ring of light changes to a spot of light of the smallest diameter that can be obtained.
9. Apply a drop of immersion oil and bring the 2 mm lens into use.

Warning. Any grease or dust in the slide, or air bubbles in the immersion oil will scatter some light and reduce the dark-ground effect.

Phase contrast microscopy

The optics of the phase contrast microscope cannot be described adequately here and reference should be made to a book on microscopy. Basically, contrast between an object and its surroundings can be obtained if the optical density of the object differs from that of the surroundings, the maximum degree of contrast being obtained if the object diffracts, or retards the light by half a wavelength compared with the direct rays. If the material to be examined is too thin, or insufficiently dense to produce a retardation of half a wavelength, it is possible to insert optical elements into the system which will increase any slight actual retardation into the desired half of a wavelength. Two elements are necessary to bring about this shift of phase, an annulus in the condenser and a phase plate which is positioned at the back focal plane of the objective. Very accurate alignment and focusing of the components is necessary and a special phase telescope replaces the eyepiece when the equipment is being set up.

Excellent phase equipment is available for the Lynx Conference microscope which a number of schools use as a microprojector. Phase kits costing about £40 are now available for some student microscopes. (Reviewed in Education in Science, Nov 1968.)

Phase contrast is the ideal method of studying live, motile bacteria, protozoans and a wide range of cytological material.

Setting up the phase contrast microscope

1. Place the slide on the stage and focus the objective and condenser as in normal microscopy. If necessary view the partially closed condenser iris to check centering. Open iris fully.
2. Remove the eyepiece and insert the phase telescope in its place. Rotate the focusing lens of the telescope until a sharp image of the dark ring of the phase plate is formed.
3. Place the annulus in the condenser.
4. Focus the condenser to give a sharp image of the bright ring of the annulus.

5. Adjust the centering of the condenser so that the bright ring of the annulus appears perfectly superimposed on the dark ring of the phase plate.
6. Remove the phase telescope and replace the normal eyepiece. Slight adjustment of the focusing may be needed.

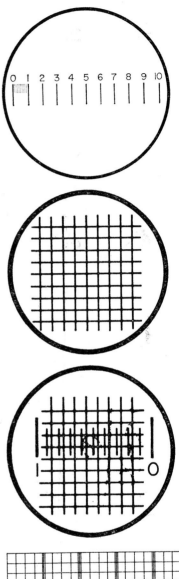

Micrometry

There are two types of measurement made with the microscope which have important applications in school experimental work. Linear dimensions can be measured by means of a graticule which has been calibrated with a stage micrometer. Number of objects in a known volume of liquid may be counted in a counting chamber such as a haemocytometer.

The graticule

This is a photographically produced grid, cemented between two glasses. It is placed on the stop, inside the eyepiece. When the top lens of the eyepiece is unscrewed to insert the graticule, care should be taken not to place fingerprints on the lens. In order to use the graticule for measurement it must be calibrated so that we know what each square represents when a particular objective is used. To calibrate the graticule, a slide with an accurate scale, usually 10×1 mm divisions with one division subdivided into 0.1 mm divisions, is placed on the stage. With this slide, or stage micrometer as it is correctly called, we can obtain a precise measurement for each division of the grid on the graticule. (Sometimes referred to as an eyepiece micrometer.)

The unit of microscopic measurement is the micrometre (preferred to micron). The correct symbol is μm (preferred to μ).

The counting chamber

A thick glass slide has four parallel grooves cut into its surface and the central platform is then ground down and polished so that its level is a certain specified distance below the general level of the slide. This part is accurately ruled into squares of known area. Many different types of rulings are used. Here we will describe the Thoma ruling.

In the Thoma ruling the ruled area occupies a square 1×1 mm, divided into 400 squares, each of side $50\,\mu$m and area of $1/400$ mm^2. Every fifth row has triple rulings as a guide. The depth of the cell is 0.1 mm. The total volume enclosed by the cell and the cover slip is 0.1 mm^3 or 10^8 μm^3 and the volume above each small square is $1/400$ mm^3 or 0.00025 mm^3.

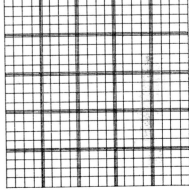

The Thoma counting chamber or haemocytometer is generally used only for counting particles smaller than about 200 μm.

Special cover slips of more than the usual thickness are used to prevent their being deformed by capillary attraction, which would lead to volumetric errors. If the concentration is low it may be necessary to make up many slides and to count them; for medium concentrations the whole of the ruled area should be counted; and for high concentrations it is necessary to dilute the sample (e.g. 1 ml diluted to 10 ml) and to multiply the result by the dilution factor to find the original concentration. The dilution should be arranged such that the whole 1×1 mm area has to be counted as the field to field variation within this sample is considerable. The variations between slides and between observers are small in comparison to the field to field variations within the slides.

A drop of the sample should be placed in the centre of the slide with a fine pipette and the cover slip placed on the slide. It is important that a standard method is adopted for traversing the haemocytometer slide. The squares should be worked from left to right and from top to bottom. **Within each square it is necessary to focus throughout the depth of the sample to locate all the particles.**

The diagram indicates which square the particles are to be included in for the purpose of counting.

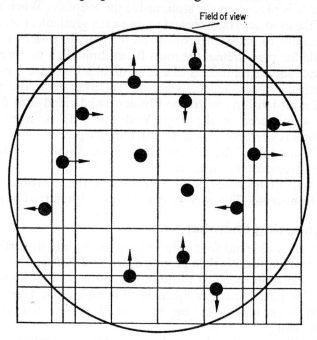

The number of particles over a given number of squares should be counted and the number per cubic millimetre of undiluted sample is obtained from the formula:

$$\frac{D \times N}{S \times K}$$

where D is the dilution, N the total number of cells counted, S the number of squares counted and K the cubic capacity above one square in mm³.

All the apparatus and particularly the slide and cover slip must be extremely clean.

For making counts of larger objects, the Fuchs-Rosenthal haemocytometer is more useful as the counting chamber is 0·2 mm deep and its base is marked with a square of side 4 mm.

MICROBIOLOGY

Precautions to be taken in microbiological laboratory work

1. Disposal of unwanted cultures

Even though pathogenic bacteria should never be used in school work, pathogens may unwittingly be grown. Bacteria are present in the air all the time and pathogens may be introduced by a sick student. Given a suitable growth medium these may then multiply to dangerous levels. Secondly some non-pathogenic bacteria may mutate to form pathogens. These forms should be avoided in school work where possible. The difficulty here is knowing which bacteria are capable of producing pathogenic mutants. It is therefore essential that all bacteria are treated as potential pathogens.

Cultures should be disposed of in the following ways:

(a) Broth cultures—in the case of non-spore forming cultures it is sufficient to remove the stopper and immerse the tube and stopper in a 10% solution of lysol. A special container should be present in the laboratory for this purpose and it should be clearly labelled.

Petri dishes should similarly be immersed in lysol.

(b) Broth cultures—in the case of spore forming cultures, the culture must be autoclaved at 121°C for 15 minutes (15 lb/in² pressure).

If spore forming organisms are to be used, a second, suitably labelled container must be present for these cultures. This container must be covered with a lid. They will then be autoclaved by the teacher. Both types of broth cultures can then be poured down the sink although it is probably better to flush them down the toilet.

(c) Solid agar cultures—these should be wrapped in paper and burnt. The Petri dishes should be placed in the 10% lysol container with the non-spore forming broth cultures.

VERY IMPORTANT. If potato slopes are to be used, some students may be tempted to experiment at home. A strong disinfectant must be applied before the cultures are disposed of.

2. The wearing of an overall is to be recommended.

3. If any bacterial culture drops on the floor or bench, etc., it must be treated as potentially dangerous, and promptly washed with 3% lysol.

4. Similarly if the hands become contaminated, they should be washed in dilute cetrimide.

5. The hands must be washed thoroughly before and after practical work.

6. Food must not be eaten in the laboratory.

14

7. Airborne particles

Whenever a liquid surface is broken a number of small liquid droplets are formed. Most of these will fall to the ground but the smallest will form an aerosol, remaining suspended in the air. If the fluid evaporates from them, a bacterial particle remains suspended in the air. Some of these particles may be as small as 1–5 μm diameter. If these are inhaled, they will rapidly penetrate to the lung.

The careless pipetting of liquids is a common source of air contamination.

When cultures are grown on solid media in screw-cap bottles, the removal of the cap may liberate airborne particles. There is often a film of fluid between the rim of the bottle and the cap. The removal of the cap breaks this film and so liberates droplets of fluid.

In laboratories regularly working with pathogens, work may be performed in safety cabinets to eliminate this risk.

8. Do not lick labels or use mouth pipettes.
9. Wash bench down with disinfectant before and after practical work.

The cultivation of micro-organisms

Culture media

Rarely can the identity of a micro-organism be recognised by its structural features alone. They must be grown in artificial culture media, and each organism must be grown in isolation from the others.

The nature of the media

(a) **Liquid or solid**—The former have two disadvantages: growths do not usually show characteristic appearances in them, and also organisms cannot be isolated in them with certainty. They are therefore of little use for identifying specific bacteria, except when used for a specific biochemical test. Solid media can be used to isolate bacteria and on this type of medium, bacteria do often show a characteristic growth habit.

(b) **Water**—tap water is unsuitable for the preparation of media. Either double glass distilled water or de-ionised water should be used in all biological work. The latter is the more convenient and time-saving. Suitable de-ionisers can be bought from the Permutit Co. Ltd, Gunnersbury Avenue, London, W.4. These have the advantage of being portable and of requiring no external source of heat or electricity. However, they may carry dissolved organic compounds derived from the resins and these affect some protozoa. Satisfactory water should have a conductivity no greater than that given by 1·5 p.p.m. of NaCl and if possible below 1·0 p.p.m. Therefore metal distilled water is unlikely to be of any use.

(c) **Agar**—this consists mainly of a long chain polysaccharide. In water solutions it sets to give a firm gel which remains unmelted at all incubation temperatures. It is decomposed or liquefied only by a few marine bacteria. It does not add to the nutritive properties of a medium and it is usually free of growth inhibitory substances.

The melting and solidifying points are not the same. Most agars melt at about 95°C and solidify at about 42°C. This means that heat sensitive nutrients can be added to it at fairly low temperatures.

After sterilising, it can be allowed to solidify for storing until use, but it should then be **remelted with as little heat as possible and it should all then be used.** Any that is still not used must not be allowed to resolidify and later heated a third time.

If the agar is made up for use in a short while it is best to keep it in a liquid state in a water bath at 45–50°C. This avoids having to spend time heating it to a high temperature.

Agar is of course absent from the liquid broth media.

(d) **Tablets or powder**—The ingredients for making up many types of media can be obtained in tablet or powder form from Oxoid Ltd, Southwark Bridge Road, London, S.E.1. The tablets are made up such that one tablet, with a stated amount of de-ionised water, will provide sufficient medium for a slope in a McCartney bottle if an agar-containing medium is used. Two tablets are required, with twice the volume of water, for a Petri dish. The tablets are more expensive, but where only small quantities are required or where no laboratory assistance is available, they are the most convenient.

The choice of media

This depends on the type of organism being grown and the nature of the investigation. If a medium is selected with a pH outside the range of most fungi, the bacteria will grow while the growth of the fungi will be inhibited. Bacteria are usually grown on nutrient agar or nutrient broth while fungi are usually grown on potato dextrose agar in stock culture. For experimental purposes other agars and broths are used. These consist of minimal media to which special reagents have been added.

16

Containers for media and cultures

If the medium is made up in large quantities, a flask with a cotton-wool plug should be used. This should also be used if a powder is being used. If tablets are being used it is more convenient to use bottles. Test tubes with cotton wool stoppers or slip-on metal caps, or screw-cap bottles can be used.

Slip-on metal caps have the advantage that they can be used repeatedly but they are no use if the cultures are to be kept for some time. Cotton wool can be used with cultures which are to be kept for some time, but they must be discarded after each use. Also if there is a change in the temperature or pressure in the tube, air will pass into it. This will be filtered as it passes through the cotton wool but this protection is not given by the slip-on caps.

Screw-cap bottles are air-tight. They also have flat bases and this has a decided advantage in the laboratory. They are supplied with rubber washers. When these are being re-used, care should be taken that no water remains between the cap and the washer.

While autoclaving screw-cap bottles, the cap must be left loose on the bottle. It must not be tightened until the bottle has cooled. If it is tightened too soon, a vacuum will be produced in the bottle and it will be difficult to remove the cap.

Sterilisation

To culture a particular type of micro-organism, we must exclude all other types by only using sterile equipment and media.

Several methods of sterilisation may be employed in schools.

1. Heat

(a) Dry heat, (b) Moist heat. Moist heat is the more efficient as the following details show:

	Times required at various temperatures	
	Moist heat	*Dry heat*
Vegetative forms of most bacteria, fungi and yeasts	10 min at 50–65°C	60 min at 100°C
Fungal and yeast spores	5–30 min at 70–90°C	60 min at 115°C
Bacterial spores	10 min at 100–121°C	60 min at 120–160°C

Moist heat kills, probably by coagulating and denaturing the proteins and enzymes, a process which requires water. Dry heat probably kills by causing a destructive oxidation of certain cell constituents.

17

The temperature and time required for killing are inversely proportional. For most purposes 15 minutes at 121°C is sufficient.

The number of micro-organisms and spores affects the rapidity of sterilisation. The number of survivors decreases exponentially with the duration of heating. The more bacteria there are present initially, the more time you should allow for sterilisation.

The nature of the material in which the organisms are heated also affects the rate of killing. Organic chemicals tend to protect organisms. Fats in particular in moist heat prevent the entry of the water into the organism.

The heat resistance of spores is greatest at neutral pH values.

2. Ultraviolet light

The effectiveness of ultraviolet light increases below 3300Å. It should not be necessary to use this as a usual technique in schools as it can damage the retina of the eye and while this damage is not permanent, it can be very painful. It also does not become apparent until some hours after contact with the source. It may however be a convenient means of performing experiments where mutations are required or of sterilising microbe contaminated, but otherwise clean plastic Petri dishes. For this purpose a lamp of the type used for examining chromatograms is the safest as the lamp can only shine down onto the bench. If it is to be used in any other way protective glasses must be worn. Ultraviolet lamps producing light of less than 3300Å units should only be used under the direct supervision of a member of staff and they should be clearly labelled to this effect.

3. Filtration

Bacteria can be removed from liquids by passing them through filters with very small pores. The bacteria do not have to be larger than the pores as other factors are also important. These factors include the physical nature and the pH of the liquid, the nature of the electrical charge on the bacteria, the pressure used and the viscosity.

Probably the most satisfactory filters are membrane filters made of cellulose acetate. These are discarded after use but they are rapid, efficient and inexpensive. They also do not retain any of the liquid being filtered. The filters are mounted on Buchner flasks. The side arm of the flask must be attached to a short length of glass tubing which contains cotton wool. This is held in place by two constrictions in the tube. It filters any air which may enter the flask and so prevents contamination of the filtrate. The Buchner flask is connected through this air filter and a non-return valve to a filter pump. The smallest negative pressure which produces filtration, and in any case not more than 25 mm Hg, should be used. Higher pressures may suck the bacteria through the filter and cause frothing of the liquids.

The filter is supported on a sintered carbon disc in a special glass or metal filter funnel. Before use the whole apparatus must be assembled, wrapped in aluminium foil, and autoclaved at 121°C for 15 minutes.

One of the most important advantages of these filters is that they can readily be used for quantitative work. The filters are about 120 μm thick and the pores in the upper surface are 0·5–1·0 μm in diameter while those in the lower surface are 3–5 μm diameter. The upper surface therefore acts as a bacterial filter while the lower surface allows nutrient media to rise through the filter by capillarity when the filter is placed on the medium. Filters with a grid marked on the upper surface to aid counting can be purchased from Oxoid Ltd.

The filters can be stored for some time.

Membrane filter apparatus can be purchased from A. Gallen-kamp Ltd.

If the membranes are handled with care they can sometimes be used again. They should be removed with fine forceps, washed in running water and steeped in a detergent solution for several hours. They should then be boiled in 3% HCl and washed again in running water, blotted and interleaved in absorbent pads.

Filters can also be obtained made of sintered glass and asbestos pads.

4. Chemicals

Few chemicals have any effect on spores. The terms disinfectant and antiseptic are usually applied to different types of substances. Disinfectants are toxic and so can only be applied to inanimate objects. Antiseptics are sufficiently non-toxic to be applied to the superficial areas of living tissue. They operate in one of the following ways:

(a) Coagulation of protein.
(b) Inactivation of enzymes.
(c) Hydrolysis.
(d) Breaking the cell membrane.
(e) Altering the permeability of the cell membrane.
(f) Forming salts with the proteins on the bacterial surface.
(g) Oxidation of sulphydryl groups from enzymes on bacterial surface.

Chemical 'sterilising' agents most commonly used in the laboratory are:

(a) **Antiseptics of the phenol group**—this includes lysol. These are mainly used for sterilising instruments, discarded cultures, and washing down the bench after cultures have been split. Usually used as a 3% solution in water. It is a protein coagulant.

(b) **Halogens**—hypochlorites are the most commonly used. They are inactivated by organic matter, and they are ineffective against spores. They are often used in a mixture with a detergent as the latter overcomes the effect of the organic matter and in particular the effect of fats.

(c) **Ethyl alcohol**—this is bactericidal at 50–70% but not at higher strengths. Its most useful application in the school laboratory is for cooling the inoculation loops after flaming when platinum loops are not being used. Other types of loops are much cheaper but they often take a few moments to cool and during this time they can become contaminated again. In this case each student should be supplied with a 3 inch × 1 inch tube containing 50% alcohol.

Sterilised instruments can be stored in alcohol.

(d) **Formaldehyde**—this is useful as it can be applied to all types of materials which would be damaged by heat treatment. It is normally used in a 0·1–0·5% solution.

(e) **Soaps and detergents**—the action is due to their concentration at the interface between water and the object, leading to a lowering of surface tension and therefore the disruption of the cells.

(f) **Cetrimide and Chloros**—both obtainable from local chemists but they are not effective against spores.

Sterilisation techniques commonly used in schools

1. Inoculating loops and points of forceps

Sterilise by holding them in the flame of the Bunsen until they are red hot.

2. Mouths of culture tubes, cotton wool stoppers, slides, cover slips, scalpels and needles

Sterilise by passing through the Bunsen flame without allowing the object to become red hot.

3. Dry Petri dishes, test tubes, flasks, pipettes, instruments and all-glass syringes

Before sterilisation, flasks and tubes must be stoppered with cotton wool. Pipettes must be wrapped in brown paper. Note that certain types of cotton wool give off volatile substances during heating which condenses on the glass. This may have to be replaced with slip-on metal caps. Sterilise in a hot air oven at 160°C for 1 hour.

Screw-cap bottles will stand the heat but the rubber washers in the caps will not. Therefore capped bottles must be autoclaved.

4. Fats, oils and greases such as petroleum jelly

Sterilise in a hot air oven. As these allow heat to penetrate only very slowly, they must be sterilised in shallow layers of no more than 0·5 cm thick in a Petri dish.

When using the hot air oven, it must not be overloaded as

spaces are required to allow air to circulate freely. At the end of the hour period **do not open the oven door for about 2 hours as glassware will crack if it is allowed to cool too quickly.**

Glassware must be absolutely dry before it is placed in the oven. Otherwise it may crack.

5. Culture media, cotton wool and other porous materials, caps containing rubber washers

Sterilise in an autoclave.

Procedure for using the autoclave

(a) Place a quantity of water in the bottom. The volume is usually recommended in the particular instruction card of the model being used.
Do not tighten the caps on screw-cap bottles.

(b) Screw down the lid, tightening diametrically opposite wingnuts in pairs so that the lid is clamped down evenly on the gasket.

(c) Open the steam cock and either light the gas or switch on the electricity.

(d) If there is a valve on the side of the autoclave for running off steam, check that it is closed.

(e) When steam issues from the steam cock, wait at least 5 minutes or whatever time the instruction card recommends, and then close the steam cock.

(f) Different autoclaves have different ways of setting the apparatus for the required temperature. The only control may be that of the gas or electric supply. On some autoclaves it may be possible to adjust the safety valve to produce the required temperature.

(g) Watch the safety valve and when steam appears from it adjust the power supply.

(h) Allow it to operate for the required time.

(i) Switch off the power supply.

(j) When the pressure gauge reaches zero, open the steam cock.

(k) After a further 5 minutes, open the lid.

If there are porous materials in the autoclave, these can be dried if there is a valve on the side of the autoclave with an outlet pipe attached to it. Before opening the steam cock, open this valve with the tube held over the sink. Water, followed by steam, still under considerable pressure, will issue from this tube. When no more steam is coming from this tube, open the steam cock to release the vacuum. This method will remove all the water from cotton wool in the form of steam.

6. Benches, old cultures, etc.

Use lysol as already described.

Sterilisation indicators

Ovens may have 'cold' spots and the one thermometer which is usually placed near the roof, is a poor guide. The pressure gauges of autoclaves often read too low. There are other useful ways of checking the effectiveness of the sterilisers.

1. Place a pellet of sulphur in a small glass tube in the steriliser. At 120°C this will change shape. It need only be exposed to this temperature for a few minutes for this change to occur.

2. Brown's tubes

These are small glass tubes containing an indicator solution. This is red until the correct time-temperature exposure is reached, when it turns green. Four types can be obtained from Brown & Co. (Leicester) Ltd. Type 2 is the most useful for ordinary laboratory autoclaving and type 3 is the most useful for hot air sterilising. They must be stored at less than 20°C to prevent deterioration.

3. Spore strips

These consist of small pieces of absorbent material soaked in a standardised suspension of Bacillus stearothermophilus spores. These spores are only killed if heated to 121°C for about 12 minutes. These strips can be bought from Oxoid Ltd in wrappers.

They should be placed in the sterilising equipment still in their wrappers. After sterilisation, remove them from their wrappers, and place in a tube of dextrose tryptone broth. Rigorous precautions must be taken against contamination while making this transfer. Incubate the tube for 7 days and then examine for growth. An unautoclaved spore strip must be cultured as a control and an uninoculated tube of culture medium as a negative control.

4. Dried earth

This can be used in envelopes in place of the spore strips as samples of earth always contain resistant spores.

There is no need to use these checks every time sterilising equipment is used, but a periodical check on the efficiency of the apparatus ought to be made.

Sterilisation of prepared media

The sterilisation time for a particular temperature is the sum of the heat penetration time, which is variable, and the holding time, which is constant for each temperature. The holding time is the time the materials must be at the desired temperature to sterilise them. The heat penetration time varies greatly with the volume of the liquid and also the type of container.

Times are as follows:

Volume of medium	Container	
	Flask	Bottle
10 ml	15 min	20 min
100 ml	20 min	25 min
500 ml	25 min	30 min
1000 ml	30 min	40 min

These are sterilisation times at 121°C.

Molten agar requires the same time as broth, but solid agar requires 5–10 minutes more.

Tubes and bottles must be placed in the autoclave so that the steam has free access to all of them. Tins must not be used as containers for the bottles unless plenty of holes are punched in the sides.

Forms in which liquid and solid media are used

Liquid media

These may be used in test tubes with slip-on metal caps or cotton wool stoppers, or in McCartney bottles with screw-on caps. 1 oz bottles are suitable if 5–10 ml is required in each bottle. For school purposes it is probably most convenient to prepare the liquids in bottles of this size, rather than prepare it in larger containers and allow the student to pour a quantity into his own bottles. This is particularly so, if tablets are being used and it should reduce the chances of contamination.

Solid media

Test tubes and bottles may be used as for liquid media. The shape in which the media are allowed to solidify depends on the type of inoculation to be used. For school purposes probably the best method is the 'slope'. This provides a large surface area for inoculation. After cooling slopes, 'water of condensation' will appear on the agar. The slopes must then be kept in a vertical position to prevent this running over the agar.

The amount of medium required for various containers is:

$6 \times \frac{5}{8}$ in test tubes require 5 ml of medium.

1 oz bottle requires 5 ml of medium.

If a large surface area is required, Petri dishes should be used. Usually 88 mm diameter plates are used. A plate this size requires 15 ml of medium. This medium is best made up and sterilised in 1 oz bottles. Each student ought to be allowed to pour his own plates. As it takes some time to melt agar, this should be done in advance of the lesson, and then the bottles of molten agar should be stored in a water bath at 45–50°C.

23

Pouring a plate

1. The agar must be molten but not hotter than can be borne on the cheek.

   If it is poured while still steaming condensation will appear on the lid. Even at this temperature (about 45°C) condensation can still occur. If the plate can be warmed slightly, it is advisable.

   The operation should be carried out away from draughts.

2. Remove the stopper from the bottle with the little finger of the working hand.

   Do not place the stopper on the bench. If you do, contamination must occur.

   Do not hold the bottle vertically, otherwise contamination will probably occur.

   Therefore do not put either the stopper or bottle down.

3. Flame the neck of the bottle.

4. With your other hand lift the lid of the Petri dish.

   Do not move the lid to the side, merely slightly raise the edge of the lid so that the medium can be poured into it.

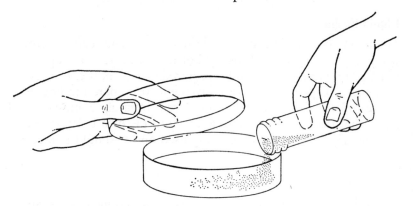

5. After replacing the lid, rotate the plate gently to ensure even distribution of the medium.

6. Leave undisturbed for $\frac{1}{4}$ hour, to allow the medium to set.

7. Store upside down to keep any condensation which may have formed off the agar.

Inoculation of culture media

A wire loop of platinum or 'nichrome' is used usually. The other end of the wire is fused into a glass rod or fastened into a metal holder. The loop is 2 mm in diameter.

The wire is sterilised by flaming in the Bunsen flame until it is red hot.

To inoculate a broth or a slope

1. If screw-cap bottles are being used loosen the caps so that they merely rest on the bottles.

2. Hold the bottle with the growth in it and the other bottle in your left hand. The tube containing the growth should be on the left.

3. Holding the inoculating wire handle in the right hand, flame the wire in the Bunsen flame.

4. If 'nichrome' wire is being used, dip the loop into a tube of sterile water or a 50% solution of alcohol to cool it.

5. Remove the cap of the tube from which the inoculation is to be made with the crooked third finger of the right hand.

6. Flame the mouth of the tube. Flaming the neck of the bottle does not necessarily kill any bacteria but it will fix them to the neck and prevent them falling into the bottle.

7. Insert loop into bottle and scrape surface of culture.

8. Withdraw the wire and remove the cap from the other bottle, holding this in the crooked fourth finger of the right hand.

9. Insert the wire charged with the growth and lightly smear the slope or place the wire in the broth.

   Withdraw the wire and flame it again.

10. Flame the necks of both tubes and replace the caps.
    If screw-caps have been used, and if they have held all the time with the washer facing downwards, it should not be necessary to flame the caps.
    It is probably best to pass cotton wool stoppers through the flame.

11. **Mark the cap with name of the bacteria and the date of inoculation.**
    Use a wax pencil for this purpose.

12. If screw-cap bottles are being used it will be necessary to tighten the caps.

This method need not be adhered to strictly, but the tubes must not be stood on the bench in a vertical position, as this increases the likelihood of contamination when the caps are removed and the caps must not be placed on the bench as contamination must occur if you do this. With practice the student will probably find that he can not develop any other efficient method which is as quick as this. The longer the inoculation takes the greater the risk of contamination.

Incubation

Most bacteria require a temperature of 37°C for optimum growth. Most fungi prefer about 25°C.

   Petri dish cultures should be incubated upside down to prevent condensation water collecting on the agar. This water would prevent the isolation of single colonies.

Stock cultures

They must be subcultured periodically to keep them alive. Having subcultured it is important to arrest growth as soon as possible in the logarithmic phase, i.e. the rapid growth in numbers which occurs soon after subculturing. This also avoids the appearance and possible selection of mutants.

The most useful media for keeping stock cultures of bacteria are egg saline, Robertson's cooked meat and litmus milk. Egg saline is probably the most useful as bacteria will survive much longer than on nutrient agar.

Two tubes should be inoculated at a time. Incubate over night at 37°C in screw-cap bottles.

Store the cultures in the refrigerator for 6–8 weeks before subculturing again.

Survival is dependent on the culture not being allowed to dry out, therefore the caps must be tight.

One of the two subcultures should be kept as the stock bottle and the other should be used for laboratory purposes.

Requirements

Per student

Loop of platinum or 'nichrome' wire in a suitable holder
Stand for the loop. This should be made of metal
Slides and cover slips
Microscope

Oil immersion lens per 2 students if possible
Wax pencil
Bunsen burner
3 in × 1 in tube of 50–70% alcohol

Per laboratory Apparatus

Cotton wool
1 oz McCartney bottles with screw caps
Flasks of various sizes for making up large amounts of culture media
88 mm diameter Petri dishes
Autoclave, preferably with a valve on the side so that steam can be evacuated at the end of the sterilisation period to dry cotton wool, etc.
Ultraviolet lamp capable of producing light of below 3300Å is useful
Cellulose acetate filters
Filter apparatus for holding the cellulose acetate filters
Buchner flask
Filter pump
Piece of glass tube, about 7 mm diameter, holding cotton wool between two constrictions
Non-return valve
Thermostatically controlled water bath
At least 3 oven/incubators. One to function as a hot air oven; one to function as an incubator at 37°C (suitable for most bacterial cultures); and one to operate at 25°C (suitable

for most fungal cultures). It is advisable to have oven/incubators rather than incubators for use at these lower temperatures only as schools can usually only afford small ones and when a large amount of materials are to be hot air sterilised, all three can be pressed into action for this
Hot water supply so that students can wash their hands whenever they require to do so
Paper towels
Phase contrast microscope
A means of rapidly measuring one volume many times, as is necessary when adding 5 ml of water to a tablet in each of many bottles. This apparatus ought to be able to measure 5, 10, or 15 ml volumes
3 buckets with lids: (a) containing 10% lysol and labelled to take unwanted broth cultures and dirty Petri dishes, (b) containing 10% lysol but labelled to take unwanted cultures which may contain spores, and (c) lined with paper towelling and labelled to take solid culture media

Reagents

Lysol
Ethyl alcohol
Formaldehyde
Sulphur
Brown's indicator tubes
Spore strips
Nutrient agar

Nutrient broth
Potato dextrose agar
Egg saline or the ingredients
Robertson's cooked meat medium
Litmus milk medium
Other agars are mentioned in connection with experimental work

The identification of bacteria

Several techniques are used to identify bacteria. Some of these are illustrated in the following pages.

1. Structure and staining reactions.
2. Cultural characters.
3. Biochemical reactions.
4. Antigenic characters.
   The serum of an animal immunised against a micro-organism contains specific antibodies which react in a characteristic manner with the particular micro-organism. An unknown bacterium may therefore be identified by demonstrating its reaction with one out of a number of known anti-sera.
5. Antibiotic sensitivity.

These tests are only valid if carried out on pure cultures. The student must therefore learn how to prepare his own pure cultures.

An experiment to demonstrate the need for various microbiological treatments

Aim

If a student is to obtain any success with his microbiological work, he must very rapidly master certain techniques. He is more likely to do this if he understands why he performs certain very simple operations, such as autoclaving rather than merely boiling to kill bacteria and the merits of cotton wool as a stopper. This experiment also demonstrates the sources of bacteria.

27

Procedure

Place 1 inch of nutrient broth into each of 12 third pint milk bottles or flasks of similar size.
Then treat these flasks as follows:

1. To demonstrate the effects of different heat treatments

Flask A. Stopper with cotton wool. Do not heat.

Flask B. Stopper with cotton wool. Heat over a boiling water bath for 10 minutes.

Flask C. Stopper with cotton wool. Heat in an autoclave at 121°C (15 lb/in² pressure) for 15 minutes.

2. To demonstrate the effects of different stoppers

Flask D. No stopper. Autoclave as for flask C.

Flask E. Stopper with cotton wool. Autoclave as for flask C.

Flask F. Stopper with cotton wool and then completely cover this with aluminium foil and securely tie this to the neck of the flask. Autoclave as for flask C.

3. To demonstrate the effects of different incubation temperatures

Autoclave unstoppered flasks as for flask C, and 4 cotton wool stoppers. Then leave them in the room after cooling for 30 minutes.

Stopper each one with sterile cotton wool.

Flask G. Incubate in the refrigerator.

Flask H. Incubate at room temperature.

Flask I. Incubate at 25°C.

Flask J. Incubate at 37°C.

4. To demonstrate the need for holding the lid over the Petri dish when subculturing

Flask K. Plug with cotton wool through which there is a piece of glass tubing of about 4 mm diameter. This should be held vertically in the cotton wool.
Autoclave as for flask C.

Flask L. Plug with cotton wool through which there is a piece of glass tubing of the same diameter as in flask K.
This tube should be bent into an S-shape.
Autoclave as for flask C.

All these flasks should then be observed daily. The date on which the first signs of turbidity appears must be noted. This marks the first signs of bacterial growth.

Continue to observe the flasks for some months, as if they have been sealed and autoclaved efficiently, some flasks will not show signs of bacteria for some considerable time.

PRECAUTION: Where the flasks are to be heated, the cotton wool stoppers must be large and fit very tightly as there is a tendency for them to be sucked through into the flask.

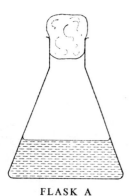

FLASK A

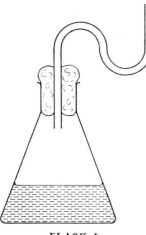

FLASK K

FLASK L

Questions

1. Why is boiling over a water bath not entirely effective?
2. Some of the autoclaved flasks show signs of bacterial growth after a time. Where do these bacteria come from?
3. How does this experiment clash with ideas on spontaneous generation?

Requirements

Apparatus

12 third pint milk bottles or flasks of similar size
1 piece of straight glass tubing, about 4 mm diameter

1 piece of S-shaped glass tubing of the same diameter as the other piece
10 cotton wool stoppers
A piece of aluminium foil, with cotton or wire

Reagents

Sufficient nutrient broth for 12 bottles
(1 inch deep in each bottle)

The isolation of pure cultures

Most studies of the physiological and other characters of bacteria are only valid when made on a pure culture, i.e. an isolated growth of a single strain free from contamination with other bacteria.

This experiment illustrates the technique of 'plating out' on a solid medium. Nutrient agar is usually used.

A loop of the culture is spread out on the agar. Where the bacteria are deposited singly at sufficient distance, all the progeny of each cell will form a discrete colony. This is readily visible to the naked eye.

Alternatively a known dilution of the culture can be used to inoculate molten agar which is then poured on to a plate.

Procedure

Method 1. Streak plate method

1. Melt two tubes of nutrient agar in a water bath.
2. Allow it to cool. This is very important; otherwise steaming would occur and the surface of the agar would get wet. This would allow the colonies to run together.
3. Remove the cap from the agar bottle and flame the neck. The bottle must be held at an angle when the cap is not in place.

29

4. Lift the lid of the Petri dish just high enough to pour the agar into it. Do not move the lid to the side.
5. Pour the second plate to serve as a control.
6. Sterilise the loop in the Bunsen flame and cool it in alcohol, and burn off the alcohol by passing the loop rapidly through the flame.
7. Remove the cap of the broth culture by grasping it in the crooked small finger of the right hand.
8. Flame the neck of the bottle.
9. Remove a loopful of culture.
10. Again flame the neck of the bottle and replace the cap.
11. Remove the lid of the Petri dish just high enough to insert the loop.
12. Smear the loop thoroughly across an area A on the plate.
13. With a free arm movement from the elbow, streak the culture back and forth over the surface of the agar, such that each streak is about $\frac{1}{4}$ in. from the next.

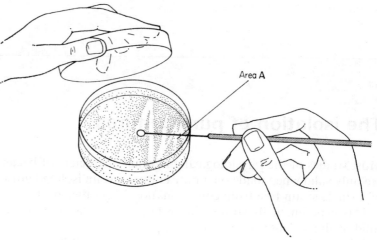

Area A

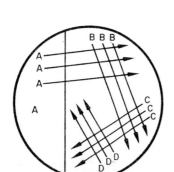

The last streaks should have thinned out the culture sufficiently to give isolated colonies.
14. In case the culture is too concentrated it is usually wise to streak out a second plate, taking the loop straight from the first plate without recharging it. This gives a greater certainty of obtaining an isolated colony.
15. A pure culture can then be obtained from an isolated colony by transferring a portion in a loop to an appropriate culture medium. (It is advisable to do a Gram stain of the colony before transferring a portion to a medium, to ensure greater certainty in selecting a pure colony.)

Method 2. Pour plate method

1. Prepare 3 tubes of nutrient broth (10 ml per tube). Number the tubes.

2. Inoculate tube 1 with 0·1 ml of the culture. Use a sterile 1 ml graduated pipette.
3. Gently shake the bottle to obtain a uniform distribution of the bacteria. Do not get bubbles in the bottle.
4. Transfer 0·1 ml of the suspension in bottle 1 to the second bottle. Use a **second** sterile pipette.

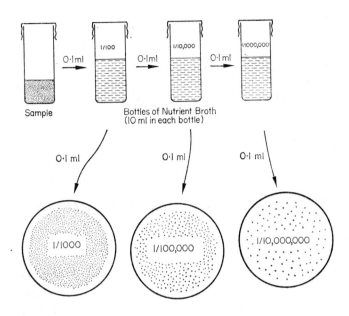

5. Shake.
6. Transfer 0·1 ml of the suspension in bottle 2 to the third bottle. Use a **third** sterile pipette.
7. Shake.
8. Mark 3 Petri dishes to correspond with the numbers on the culture dilution bottles.
9. Transfer 0·1 ml of each culture dilution to the appropriate Petri dish.
10. Melt 3 tubes of nutrient agar. Allow to cool, and pour into the Petri dishes.
11. Gently rotate the plates to obtain a uniform distribution of organisms.
12. When the agar has set, invert the plates and incubate.

The first agar plate usually contains far too many colonies with the result that it is difficult to find a well-isolated one, but the second and third plate should provide an isolated colony. Many of the colonies will be deep in the agar and therefore it is usually easier to obtain an isolated colony by this method.

PRECAUTION: In both these methods it may be necessary to use a mounted needle as a means of removing part of a colony. A loop may remove part of an adjacent colony.

31

Requirements
Streak plate method

Inoculating loop and needle
2 previously sterilised bottles of nutrient agar
2 sterile Petri dishes
3 in × 1 in tube of alcohol

Water bath
2 sterilised slopes or plates of any agar
Gram stain
Bunsen burner

Pour plate method

3 sterilised bottles of nutrient broth, each containing 10 ml
3 sterile graduated 1 ml pipettes, each still wrapped in brown paper
3 sterile Petri dishes

Inoculating loop and needle
Bunsen burner
Gram stain
3 sterile tubes of nutrient agar, 15 ml in each

The examination of living bacteria

Procedure

Method 1. Use the 'hanging drop' method to observe motility

1. Apply Vaseline around the depression of a cavity slide.
2. Using an inoculating loop, aseptically transfer one drop of the liquid bacterial culture to the centre of a clean $\frac{5}{8}$ in square No. 1 coverslip.
3. Rest the coverslip on a tile and place the cavity slide on the coverslip, centered over the drop of culture.
4. Quickly and carefully turn the slide the right side up. The hanging drop should be suspended in the well of the slide.
5. To examine the drop, first locate its edge in the centre of the microscope field with the low-power objective and considerably reduce the light. You will see the edge as a bright wavy line against a grey background. Turn to the high-power objective and refocus on the edge of the drop. The bacteria in the drop will then be in approximate focus and you can detect them by their movement and their contrast to the surrounding medium. Non-motile bacteria will exhibit Brownian movement. Motile bacteria usually move for considerable distances in one direction.

Cultures for motility

1. To isolate motile bacteria, use a Craigie tube. This consists of a 1 oz McCartney bottle containing a piece of glass tubing 5 cm × 5–6 mm.
2. To the tube add 12 ml of sloppy agar (broth containing 0·1–0·2% agar).

32

3. Place the glass tube in the bottle. There must be sufficient clearance between the top of the tube and the surface of the medium so that a meniscus bridge does not form. The connection between the inner tube contents and the agar in the bottle should be through the bottom.
4. Inoculate the inner tube.
   The sloppy agar does not permit bacterial movements in convection currents and so the only bacteria which can move to the outer bottle will be the motile ones.
5. Incubate and subculture from the outer bottle.

Method 2. Wet film technique

Place a drop of the culture on a clean microscope slide.
Place a coverslip on the drop.
Examine with the dark ground microscope.

Requirements
Method 1. Hanging drop method

Cavity slide
Vaseline
⅝ in square No. 1 cover glasses
Cultures for motility
(*Serratia marcescens*)

1 oz McCartney bottle containing a piece of glass tubing
5 cm × 5–6 mm
12 ml of sloppy agar (broth containing 0·1–0·2% agar)
Inoculating loop

Method 2. Wet film method

Microscope slide
Coverslip
Dark ground microscope

Preparation of a stained smear from a broth or slope

Procedure

Making a fixed smear

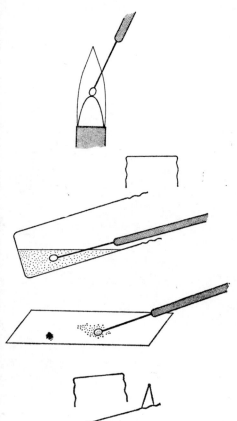

1. With the sterilised loop, place a small drop of the broth on a clean slide.

   If the culture is taken from a solid medium, first place a drop of sterile water on the slide, and thoroughly mix the bacterial material with it.

2. Gently stir the loop in the film on the slide to obtain an even distribution of cells over an area about the size of a sixpence.

3. Flame neck of bottle from which culture has been taken and replace the cap.

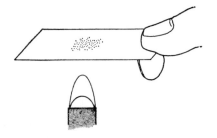

4. Air dry the slide, or hold it high above the Bunsen flame.

5. Heat fix the smear when it is dry by passing it, smear upper-most, through the flame three or four times.

PRECAUTION: Too much heat will distort the shape of the cells. The slide should merely feel warm to the back of the hand.

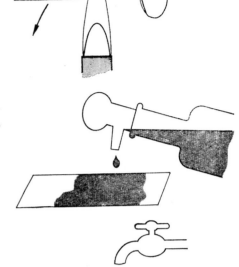

Simple staining with basic dyes

1. Flood the smear with stain and allow time for the dye to act. Allow $\frac{1}{2}$ minute in carbol fuchsin, 1 minute in crystal violet, 1 minute in safranin, and 3–5 minutes in methylene blue.

2. Wash the slide gently in running water over the sink.

3. Dry the slide between layers of blotting paper.

4. Examine the slide, first with low power, choosing a suitable area for oil immersion examination.

Observations

Note the following features:

1. Size of the cells.
2. Shape of the cells.
3. Grouping of the cells.
4. If you have too heavy a smear, you will have conglomerations of cells.
5. The density of the stain in the cells. If this is not satisfactory, try again, adjusting the time in the stain.

Requirements

Inoculating loop
Microscope slides
Bunsen burner
Bacterial culture, a sample of saliva will do
Stain, any one of the following will

do: crystal violet, safranin, carbol fuchsin, or methylene blue
Blotting paper
Oil immersion objective and immer-sion oil
Microscope

Differential staining

Simple staining depends on the fact that bacteria differ from their surroundings chemically and so can be stained to contrast with their surroundings.

Differential staining makes use of the fact that bacteria differ from one another chemically and physically to distinguish between different types of bacteria. They will react differently to one staining procedure.

An example of a differential stain is the Gram stain. This divides bacteria into two groups, those which stain purple and those which stain pink. The former are Gram-positive and the latter Gram-negative.

Gram staining involves treating the cells in four ways:

1. Staining with a basic dye as in the last experiment.
2. Treating with a mordant, i.e. a chemical which increases the affinity of the cells for the dye.
3. Decolourising agent. Some cells decolourise more readily than others and it is this property which is used to distinguish cells when Gram staining.
4. Counter stain, i.e. a stain of a different colour to the first one. This stains the decolourised cells a different colour from the cells which were not decolourised.

Procedure

1. Fix a smear of *Escherichia coli, Sarcina lutea and Staphylococcus albus.*
2. Stain with crystal violet for 1 minute.
3. Wash in gently running water.
4. Cover the smear with Gram's iodine solution.
5. Allow this 1 minute to act.
6. Drain off the excess iodine and rinse with running water.
7. Decolourise by washing with 95% alcohol. Hold the slide on a slight slope and drip the alcohol on to the slide. Continue doing this until the drops coming off the slide are a pale violet colour.
8. Wash in gently running water.
9. Counter stain with safranin for 1 minute.
10. Wash in gently running water.
11. Dry between layers of blotting paper.
12. Examine under the oil immersion lens.

Requirements

Cultures of *E. coli, Sarcina lutea*, and *Staphylococcus albus*
Stains: crystal violet, and safranin
Gram's iodine

95% ethyl alcohol in a small beak bottle
Blotting paper. Microscope, immersion oil and oil immersion lens

The staining of endospores

These are bodies produced in the cells of many bacterial species. They are more favourable to unfavourable conditions than the vegetative cells producing them. It is doubtful whether such extreme conditions ever occur in nature, and therefore whether endospores ever naturally occur. There seems to be a close relationship between spore formation and the exhaustion of the nutrients needed for the continued vegetative growth.

Conditions which have been reported as favouring sporulation are the addition of salts of manganese, chromium and nickel to the growth medium, the shaking of a culture in de-ionised water at 37°C, addition of tomato juice to the medium, keeping the pH at 5·5 or above, and the addition of certain amino acids to the medium. The conditions necessary for sporulation in one species do not necessarily apply for another. No guide can be given here for producing sporulation with certainty.

If spore-bearing bacteria are stained with basic dyes or with Gram's stain, the body of the cell is deeply stained but the spore not stained. It appears as a clear area in the cell.

With more vigorous staining techniques, dye can be introduced into the substance of the spore. Decolourising reagents then can be used to remove the stain from the rest of the cell.

Procedure

Method 1

1. Stain with carbol fuchsin for 3–5 minutes, heating the preparation until steam rises from it.
2. Wash in de-ionised water.
3. Treat with $\frac{1}{2}\%$ sulphuric acid for one to several minutes. (Determine the time by trial and error.)
4. Wash in water.
5. Counterstain with 1% aqueous methylene blue for 3 minutes.
6. Wash in water and blot dry.

The spores are stained red and the protoplasm of the bacilli blue.

Method 2

As above but instead of counterstaining with methylene blue, thinly spread a drop of 10% nigrosin over the dried decolourised smear with another microscope slide. This gives a dark background which outlines the unstained bacteria.

37

Method 3

1. Place the dried, fixed smear over a beaker of boiling water with the smear uppermost.
2. When condensation appears on the underside of the slide, flood the smear with a 5% solution of malachite green. Leave this on for 1 minute.
3. Wash in cold water.
4. Treat with 0·5% safranin for $\frac{1}{2}$ minute.
5. Wash and dry.

The spores stain green and the vegetative bacilli red. Examine cells stained with these methods with the oil immersion lens.

Requirements

Cultures of *Bacillus subtilis*	5% malachite green
Stains:	0·5% safranin
carbol fuchsin, Ziehl-Neilsen's formula	0·5% sulphuric acid
	Blotting paper
1% aqueous methylene blue	Microscope, immersion oil and oil
10% nigrosin solution	immersion lens

Observations on the appearance of the colony

Observe the following colonial characteristics: size of colony, its shape in plan view and elevation, translucence and opacity, pigmentation.

Also note any changes in the colour and consistency of the medium. These observations should be made on plates. Also

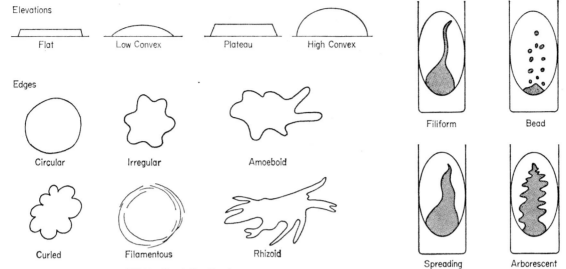

Surface: Smooth, Glistening, Wrinkled, Rough, Dry, Powdery

observe the growth on an agar slope which has been streaked with the culture.

Differential plating

This means of identifying bacteria makes use of their physiology. Some bacteria will ferment lactose, therefore by incorporating a pH indicator into the agar it is possible to detect the bacteria visually.

Procedure

1. Prepare 3 tubes of lactose yeast extract agar containing bromo-cresol purple indicator (10 ml per tube).
   Number the tubes. Keep in a water bath at 45°C until pouring.
2. Inoculate tube 1 with 0·1 ml of milk. Use a sterile graduated 1 ml pipette.
3. Gently shake the bottle to obtain a uniform distribution of bacteria. Do not get bubbles in the bottle.
4. Transfer 0·1 ml of the suspension from the first tube to tube 2 using a **second** sterile pipette.
5. Continue until the three tubes have been inoculated in this way. (See the pour plate method, on p. 30.)
6. Mark three Petri dishes to correspond with the numbers on the tubes.
7. Pour each of the tubes of agar into the appropriately labelled plate. Rotate each plate to distribute the agar evenly.
8. Allow to cool and incubate at 37°C until the next period.

Observations

Examine the plates for colonies surrounded by yellow zones. A yellow area indicates the presence of acid as the indicator becomes yellow below pH 5·1.

Requirements

Apparatus

3 McCartney bottles
Water bath at 45°C and incubator at 37°C

3 sterile graduated 1 ml pipettes
3 sterile Petri plates

Reagents

Milk
10 ml of lactose yeast extract agar containing bromocresol purple in each of the McCartney bottles. These should be sterilised
Add 10 g tryptone, 10 g yeast extract, 5 g K_2HPO_4, 5 g lactose, and 15 g of

agar to 1 litre of sterile water. This should have a pH of 7·0
To make up the indicator, add 16 g to 500 ml of 95% ethyl alcohol and 500 ml of sterile water. Filter
Add 2 ml of this indicator to 1 litre of medium

Selective plating

It is possible to select some types of micro-organisms from an impure sample by adding a chemical to the medium which will repress the growth of some of the bacteria.

Sodium azide poisons iron-containing, oxidative systems and so represses bacteria containing such systems. Other bacteria are relatively resistant to this poison. Endo agar containing basic fuchsin suppresses the Gram-positive species.

Procedure

1. Prepare 3 tubes of tryptone yeast extract agar containing 0·02% sodium azide and 3 tubes of Endo agar (10 ml per tube). Number each of the tubes. Keep in a water bath at 45°C.
2. Prepare a suspension of rat or mice faeces in sterile water.
3. Inoculate tube 1 of each of the agars with 0·1 ml of this suspension. Use a sterile graduated 1 ml pipette.
4. Continue as in the pour plate method and the differential plate method.
5. Incubate the plates at 37°C for 2 days.
6. After 2 days inoculate a bottle of tryptone yeast extract broth with a typical colony from the azide agar and a second bottle of broth with a typical colony from the Endo agar.
7. Incubate the tubes at 37°C for two days.
8. After 2 days prepare Gram stain preparations.

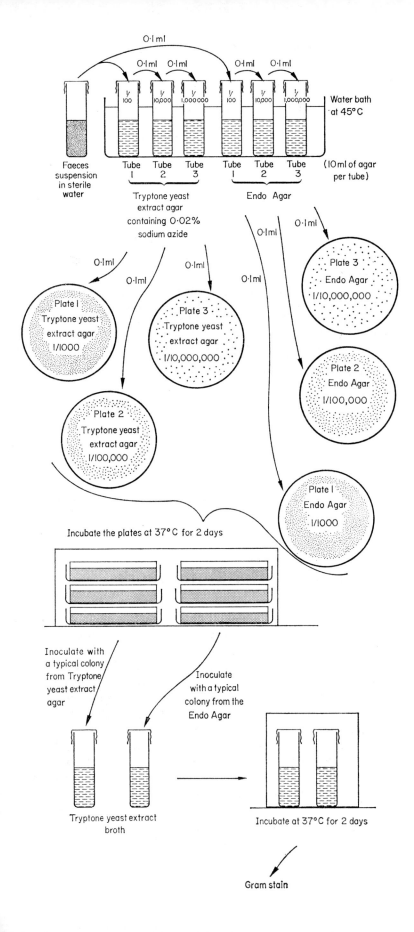

0·1ml

0·1ml   0·1ml   0·1ml   0·1ml

1/100   1/10,000   1/1,000,000   1/100   1/10,000   1/1,000,000   Water bath
at 45°C

Faeces   Tube   Tube   Tube   Tube   Tube   Tube   (10ml of agar
suspension   1   2   3   1   2   3   per tube)
in sterile
water   Tryptone yeast   Endo Agar
           extract agar
           containing 0·02%
           sodium azide

0·1ml   0·1ml

0·1ml   0·1ml   Plate 3
                                                      Endo Agar
                          0·1ml   1/10,000,000

Plate 1   Plate 3
Tryptone yeast   Tryptone yeast
extract agar   extract agar
1/1000   1/10,000,000   Plate 2
                                                      Endo Agar
                                                      1/100,000

Plate 2
Tryptone yeast
extract agar
1/100,000   Plate 1
                                             Endo Agar
                                             1/1000

Incubate the plates at 37°C for 2 days

Inoculate with
a typical colony
from Tryptone   Inoculate
yeast extract   with a typical
agar   colony from the
                        Endo Agar

Tryptone yeast extract   Incubate at 37°C for 2 days
broth

Gram stain

Observations

1. Note the differences in the colonies which develop on each of the agars.
2. Note the morphology of the bacteria in each of the Gram stained preparations.

Requirements

Apparatus

8 McCartney bottles
Water bath at 45°C and an incubator at 37°C

6 sterile graduated 1 ml pipettes
6 sterile Petri dishes

Reagents

10 ml of Endo agar in each of 3 of the McCartney bottles
These should then be sterilised
To make this up add 10 g of lactose, 10 g of peptone, 3·5 g of K_2HPO_4, 2·5 g of sodium sulphite, 0·5 g of basic fuchsin and 15 g of agar to 1 litre of sterile water

To make up the tryptone yeast extract agar, add 5 g of yeast extract, 10 g of tryptone, 5 g of K_2HPO_4, 1 g of glucose and 15 g of agar, to 1 litre of sterile water. Add 0·2 g of sodium azide
Gram stain

Biological

Rat or mice faeces

Estimating bacterial numbers

Counts may be made as a **total count** or as a **viable count**. The former includes all bacteria, dead and alive; the latter only live bacteria.

The viable count assumes that a viable growth will occur from each organism, but bacteria are rarely separated completely from each other. Therefore one colony may develop from several bacteria. It is also rare for bacteria to be distributed evenly through a sample. As only small samples are usually examined, large errors are possible. It is therefore essential that several plates of each dilution are prepared. A liberal interpretation of results is required.

The culture medium may not permit some of the bacteria to grow, or the incubation temperature may be unsuitable for the growth of some of the bacteria.

Total counts are made using physical methods, such as direct counting, and measuring the turbidity of the medium.

Viable counts involve serial dilutions of the bacterial medium followed by the plating out of aliquots of each dilution. Each colony growing is then assumed to have originated from one viable unit, i.e. one organism or many.

The counting chamber method for making total counts

For details of counting chambers see the section on micrometry (page 9).

Procedure

1. Add 2–3 drops of 40% formaldehyde to 10 ml of the suspension to fix it. Mix this thoroughly.
2. Place a small drop or loopful on the centre of the chamber platform and place the coverslip over it.
3. Press the coverslip down until coloured Newton's rings are seen clearly over the areas of contact.
4. If a phase contrast microscope is available, examine the bacteria with this. Alternatively use an ordinary microscope with the iris slightly closed and with the condenser slightly out of focus.
   It may be found to be necessary to stain the preparation with methylene blue. If so, a second slide will have to be prepared. Add the dye to a concentration of 0·1%. Use the 4 mm objective.
5. Count sufficient squares to include several hundred bacteria. Focus at different levels in each square so as to include all the bacteria in each square.
   Select the square in a prearranged way, i.e. every fifth square for instance.
6. Determine the average number of bacteria per square.
7. To obtain the count per ml in the original suspension, multiply the average number of bacteria per square by 4,000,000 and by the dilution factor.
8. Repeat with two more preparations and take the average of the three results.
9. If the original suspension contains much less than 40,000,000 bacteria per ml a counting chamber with a 0·1 ml chamber should be used instead of the one with a 1 mm side.

The pour plate method for making viable counts

See the section on the pour plate method (page 30).
A measured amount of the suspension is mixed with molten agar and after setting and being incubated, the number of colonies is counted.

Procedure
1. Prepare 4 tubes of nutrient agar (15 ml per tube).
   Then keep them at 45°C in a water bath.
2. Using a sterile graduated 1 ml pipette, add 1 ml of the bacterial suspension to 9 ml of sterile water in a McCartney bottle.
   Transfer 0·1 ml from the remaining suspension into a sterile Petri dish. Clearly label this as containing the original suspension diluted 1 in 10.
3. Vigorously shake the McCartney bottle to ensure the thorough mixing of the bacteria. This should also break apart bacteria which tend to clump together.

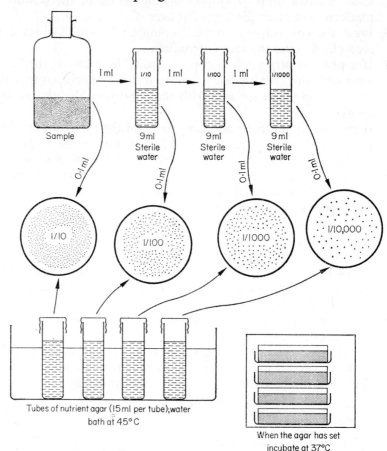

Tubes of nutrient agar (15ml per tube);water bath at 45°C

When the agar has set incubate at 37°C

This tube now contains a 1 in 10 dilution of the original suspension.

4. **Using a second sterile pipette**, pipette 0·1 ml from the 1 in 10 dilution into a second sterile Petri dish, and 1 ml into 9 ml of sterile water. Label the dish as containing a 1 in a 100 dilution of the original suspension.

   The McCartney bottle will also contain a 1 in 100 dilution. Shake the bottle as above.

5. Repeat the procedure **using a third sterile pipette.**

   This will give dilutions of 1 in 1,000 in both a third Petri dish and in a third McCartney bottle.

6. Pour the melted cooled agar into the Petri dishes and mix the agar thoroughly with the bacteria.

7. Gently rotate the plates to obtain a uniform distribution of bacteria.

8. When the agar has set incubate at 37°C in an inverted position.

Results

After 2–3 days count the number of colonies which have appeared. To avoid counting a colony twice, mark each colony on the plate with a wax pencil. It is best to count the colonies with the plate in the inverted position.

As already stated, several plates of each dilution must be prepared, therefore the results of the whole class must be pooled.

In determining the average it must be remembered that bacteria are usually distributed very unevenly throughout a culture medium, it is therefore best to make use of Simpson's rule when determining the average.

Consider an odd number of plate counts, and arrange the results in ascending order.

Add the first result to 4 times the second result.

Add this figure to 2 times the third result.

Add this figure to 4 times the fourth result.

Add this figure to 2 times the fifth result.

Continue treating these results in this way but the last but one result must be multiplied by 4 and the last result by 1. Divide the total by sum of the fours, twos and ones.

This method of determining the average is giving less weight to the extreme results than to the other results.

Requirements

Apparatus

Counting chamber
Phase contrast microscope if available, or an ordinary microscope with an iris and a condenser
6 McCartney bottles, 3 containing 15 ml of nutrient agar (sterile)
Water bath at 45°C
3 sterile graduated pipettes (1 ml).

These should be sterilised in brown paper and kept in it until being used
10 ml sterile graduated pipette (for transferring 9 ml of sterile water in to each of three McCartney bottles)
3 Petri dishes (sterile)
Incubator at 37°C

Reagents

40% formaldehyde

0·1% methylene blue

Nutrient agar, 15 ml in each of 3 McCartney bottles

The cultivation and examination of fungi

Cultivation media

Media which select for fungi are inhibitory to bacteria. Fungi prefer acid media. Note that agars are hydrolysed at low pH values by heat. These media must not therefore be autoclaved above 115°C. After autoclaving, the medium may be allowed to solidify in bulk but it should only be reheated once, and then only with a minimum of heat.

Overheating should also be avoided as many of these agars contain a considerable amount of sugar which tends to char.

Some useful agars are:

(a) **Sabourard's agar.** The low pH value makes this very useful for selecting against bacteria.

(b) **Malt agar.** The high sugar content makes this particularly suitable for saprophytic as well as parasitic yeasts and fungi. This is the best medium for highly exacting fungi.

(c) **Potato dextrose agar.**

These agars can be purchased from Oxoid Ltd.

Slide culture of fungi

Method 1

This is a useful means of studying the growth and sporulation of fungi without disturbing them.

1. Sterilise a Petri dish containing a lining of filter paper. This plate can be sterilised in an autoclave as the paper must be moist. Place glass rods on the paper to support the slide.

2. Dip a clean glass slide in alcohol and pass it through a flame to burn off the alcohol.

3. Place the slide on the glass rods on the filter paper.

4. Using a sterile pipette, transfer a small quantity of sterile agar to the centre of the slide.

5. The agar can then be inoculated either in the centre or one edge of the agar can be cut straight with a sterile scalpel and this edge inoculated with the fungus.

6. Place a sterile coverslip over the culture and replace the Petri dish lid.

7. When the fungus has grown sufficiently, remove the coverslip.

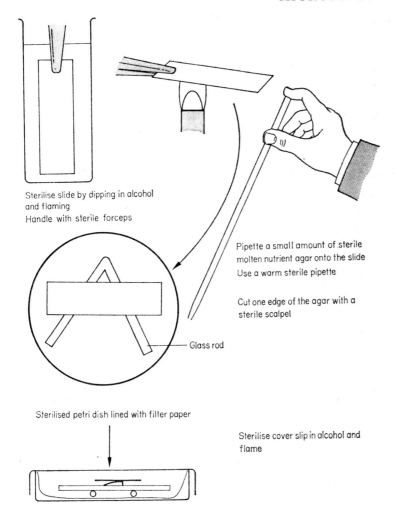

Sterilise slide by dipping in alcohol
and flaming
Handle with sterile forceps

Pipette a small amount of sterile
molten nutrient agar onto the slide
Use a warm sterile pipette

Cut one edge of the agar with a
sterile scalpel

Glass rod

Sterilised petri dish lined with filter paper

Sterilise cover slip in alcohol and
flame

During incubation at 25°C keep the filter paper moist with sterile water.

The fungus usually adheres to the coverslip or the slide.

8. If the fungus is on the coverslip, place a drop of lactophenol on the fungus and place the slip on a clean slide. If the fungus is adhering to the slide, remove the agar and place a drop of lactophenol on the fungus. Cover with a clean coverslip.

Method 2. (After Dade & Gunnell 'Classwork with Fungi', Commonwealth Mycological Institute.)

This method is designed to obtain the growth in the centre of the coverslip rather than near its edge as in the previous method.

1. Pour a plate of agar and allow it to set.
2. With a sterile scalpel, cut two slits in the agar at right angles to each other.
3. Lift each of the triangular lips formed by the cuts in turn and place a sterile coverslip under the lip.

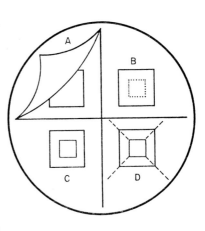

47

4. Invert the dish and mark the position of the coverslips with a wax pencil by a small square over the centre of each slip.
5. Cut out squares of agar above the wax marked squares.
6. Inoculate the plate in the centre.
7. Incubate at 25°C.
8. When the growth has covered the coverslips, these may be removed from the plate and mounted in the usual way. Prior to this they can be examined while still on the plate under the binocular microscope.

Requirements

Apparatus

Sterile Petri dishes
Slides and coverslips
Microscopes and binocular micro-scopes
Filter paper
Bent glass rods

Reagents

Sabourards agar
Malt agar
Potato dextrose agar
Lactophenol

The production of vitamins by fungi

Some fungi are self sufficient for vitamins while others are partially deficient for some vitamins. For instance *Aspergillus rugulosus* secretes biotin into the culture medium while *Sordaria fimicola* fails to produce fertile perithecia in the absence of biotin.

Procedure
1. Prepare a Petri dish of the vitamin deficient medium.
2. Inoculate a point on the medium with *A. rugulosus* and then inoculate a second point about 2 cm away with *S. fimicola*.
3. Set up a control culture of *S. fimicola* only.

Results
Observe the position of the perithecia

Requirements
Apparatus
Petri dish previously sterilised

Reagents
Culture medium made up thus: glucose 5 g, KNO_3 3·5 g

$MgSO_4 \cdot 7H_2O$ 0·75 g, KH_2PO_4 1·75 g, agar 20 g, de-ionised water 1 litre

Biological
Aspergillus rugulosus

Sordaria fimicola

The production of antibiotics by fungi

Procedure
1. Pour a plate of nutrient agar and allow it to set.
2. Make a suspension of *Penicillium notatum* spores in a drop of sterile water, on a microscope slide.
3. Make a smear of this suspension across the plate near one edge.
4. Three days later inoculate the plate with bacteria, *Serratia marcescens* and *Sarcina lutea*. Smear the agar at right angles to the fungal smear. The two bacterial smears should be about 2 cm apart and should be right up to the fungal smear. In smearing move towards the fungal smear.
5. Incubate at 37°C for 2 days.

Results
Observe the extent of the growth of both bacteria

Requirements
Apparatus
Sterile Petri dish of nutrient agar   Incubator at 37°C
Inoculating loop

Reagents
Nutrient agar

Biological
Penicillium notatum   *Sarcina lutea*
Serratia marcescens

The use of filter paper discs for determining the sensitivity of bacteria to antibiotics

Small discs of filter paper are impregnated with antibiotic. 'Multodisks' are usually used with several arms, each impregnated with a different antibiotic. There are numerous advantages to using the 'Multodisk' rather than single pieces of filter paper. They are labour saving; they reduce the chance of contamination; and as all the discs are attached to one piece of filter paper, when placed on the agar, the moisture content will be the same in each disc. Where separate discs are used differences in the moisture content of the discs may cause differences in the rate of diffusion of the antibiotic out of the disc.

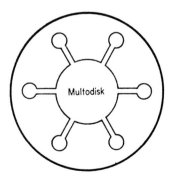

49

Procedure

1. Melt a tube of nutrient agar.
2. When this has cooled so that it can comfortably be touched with the hand, inoculate it with a loop of *E. coli*.
3. Pour the agar into a sterile Petri plate.
4. Allow the agar to solidify, and then transfer a 'Multodisk' with flamed forceps to the surface of the agar.
5. Incubate the plate at 37°C.

Results

Observe zones of inhibition. The size of the zones is not necessarily an indication of the susceptibility of the organism to the antibiotics.

Requirements

Apparatus

McCartney bottle
Sterile Petri plate

Forceps
Incubator at 37°C

Reagents

'Multodisk' This can be purchased from Oxoid Ltd
Note that most reagents used in 'Multodisks' have a shelf life of only 12 months and that penicillin should only be kept for 6 months
When bought the 'Multodisks' are in

air-tight tins containing a quantity of silica gel drying agent. They may be stored at room temperature but once opened they should be kept in the refrigerator
Nutrient agar in the McCartney bottle, sterilised

Biological

E. coli

Experiment to distinguish the bactericidal action of antibiotics from their bacteriostatic actions

A zone of inhibition around a disc of antibiotics does not necessarily mean that it is killing all the bacteria in that zone. It may merely be inhibiting the growth of cells in the zone. The presence or absence of living organisms in this zone may be shown by the replica plate method. About 1% of the organisms are transferred by this method.

Procedure

1. Pour a plate of nutrient agar and allow it to solidify.
2. On the bottom of the plate draw an outline of the 'Multodisk' used in the previous experiment with a wax pencil.
   Indicate the nature of the antibiotic in each of the small discs.
3. Press the velvet face of the 'stamp' on to the surface of the plate from the previous experiment. Avoid lateral movement.
4. Lift the 'stamp' off and press it firmly on the replica plate. Take care that it is placed on the replica plate, orientated in the same manner as on the first plate.
5. Incubate the replica plate at 37°C and examine for growths.

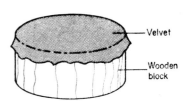

Requirements

Apparatus

Sterile Petri dish of nutrient agar
Petri dish containing antibiotic from previous experiment
Stamp. This can be made by cutting a cylindrical wooden block, about 3 cm high and slightly less than the diameter of the Petri dish. The surface must be even and smooth.

Paint one face with an adhesive and stick a piece of velvet to it. Trim this to fit the face of the wood. The pile of the velvet must face away from the wood
The 'stamp' must be sterilised. This can be done in an Autoclave
Incubator at 37°C

The measurement of growth

1. Colony growth

Procedure

1. Sterilise a microscope slide in a Petri dish.
2. Pour a tube of nutrient agar aseptically onto the slide.
3. When the agar has set, inoculate the agar with a small amount of a bacterial colony.
4. Dip a long coverslip in alcohol and pass it through the Bunsen flame. Allow it a moment to cool and place it over the culture.
5. Replace the lid of the Petri dish.
6. Observations must be made daily or more often if possible. Use the low-power objective to mark the point to which the colony has grown.
7. Mark each end of the colony with a wax pencil at each observation.
8. Measure each days growth in mm.

NOTE. This method can be used with fungi but potato dextrose agar should be used.

Results

Draw a graph of size of colony against time.

Questions

1. Why may this method give a misleading result?
2. What effect will the concentration of the media have on the results?

Requirements

Apparatus

Sterilised Petri dish containing a microscope slide

Inoculation loop

Long coverslip

Wax pencil

Reagents

McCartney bottle of nutrient agar or potato dextrose agar

Alcohol

Biological

Culture of bacteria or fungi

---

2. An alternative method for measuring colony growth

Procedure

1. Draw two lines at right angles to each other and across the middle of the base of a sterilised Petri dish.
2. Pour a bottle of sterile potato dextrose agar onto the plate.
3. When this has set, inoculate it with a fungus. The inoculum should be a disc of agar cut from the edge of a growing colony with a sterilised cork borer. Place it on the agar at the intersection of the lines on the Petri dish.
4. At regular intervals measure the extent of the colony along each of the diameters.

Results

Plot a graph of the size of the colony against time.

Requirements

Apparatus

Sterile Petri dish

Wax pencil

Reagents
Bottle of potato dextrose agar

Biological
Culture of fungus

---

3. The direct measurement of hyphal growth

Procedure
1. Inoculate a Petri dish of potato dextrose agar with a suitable fungus.
2. Incubate the plate at 25°C.
3. When some growth has taken place, remove the lid of the plate and examine the plate under the low power objective of the microscope.
4. Select one hyphal tip near the edge of the growth.
5. Insert a graticule in the eyepiece of the microscope.
6. Rotate the plate until the hypha you have focused on is parallel with the scale and its tip in line with one of the marks.
7. Observe the position of the hypha tip as often as possible. Considerable growth usually occurs in a few hours. Therefore several observations should be made each day.
8. Calibrate the graticule with a stage micrometer.
9. Convert your observations of hyphal length on the graticule into actual units of length.

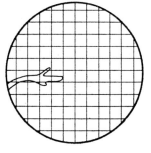

NOTE. It is best to leave the plate on the microscope stage throughout the experiment. As it lasts for such a short time, contaminants will not have time to grow.

Results
Plot a graph of hyphal length against time.

Question
What are the limitations of this method?

Requirements
Apparatus
Sterile Petri dish containing a thin layer of agar

Graticule and stage micrometer

Reagents

Potato dextrose agar

Biological

Fungus such as *Sordaria* or *Mucor*

---

4. Growth as measured by increases in dry weight

Procedure

1. Pour an equal volume of nutrient broth into each of ten 250 ml conical flasks.
   Plug the flasks with cotton wool.
2. Inoculate each in the centre with a small disc of fungal culture on agar. There must be as little agar as possible.
3. At intervals, gently rotate the flasks to aerate the medium.
4. After 2 days harvest the growth in the first flask. To do this filter the growth and medium through fine muslin or a piece of nylon stocking.
5. Wash the residue several times with de-ionised water.
6. Transfer the clean residue to a McCartney bottle cap which has been previously weighed.
7. Dry the sample in an oven until it is a constant weight. The temperature for this should be about 95°C.
8. Cool, prior to weighing, in a desiccator containing calcium chloride.
9. Two days later harvest the growth in the second flask.
10. Now repeat the weighing as for the first flask.
11. At two day intervals, repeat the procedure with the growths in the other flasks.

Results

Plot the graph of the dry weight of the growths against the time.

Requirements

Apparatus

Ten 250 ml conical flasks, each sterilised
Microbiological loops
10 pieces of fine muslin, or 10 pieces of nylon stocking

10 McCartney bottle caps
Oven at 95°C
Desiccator containing CaCl₂

Reagents

Nutrient broth. De-ionised water

Biological

Culture on nutrient agar of *Sordaria*
or a similar fungus

---

5. The turbidimetric estimation of growth

The turbidity of a suspension is due to the light scattered by the particles in the suspension during its passage through the suspension. A photocolorimeter can be used to measure the light lost from the beam due to scattering, and hence the amount of turbidity (See appendix 10).

If the turbidity of a series of bacterial cultures of known concentrations are plotted against the actual counts, a calibration curve is obtained from which it is possible to determine the concentrations of other bacterial suspensions if the turbidity is known.

This calibration curve applies only to the one particular organism and only under the one set of conditions. A new curve must be prepared for each new organism under each new condition. A change in shape of the cell will change the turbidity.

Note that the growth medium can only be used as the blank if the amount of light it absorbs is not altered by the growth of the organisms. If it is altered, the cells must be washed and resuspended in fresh medium before being placed in the colorimeter.

Low concentrations should always be used because this gives a linear calibration, whereas high concentrations give a curve.

Procedure

Preparation of calibration plot

1. Make a plate count of the culture of bacteria provided using the pour plate method. (See page 44 for details.)
   Use the following dilutions, 10^{-3}, 10^{-4}, 10^{-5}, 10^{-6}, 10^{-7}, and 10^{-8}.
2. The plates should then be placed in the incubator at 37°C.
3. Insert a tube of sterile nutrient broth in the colorimeter and set the instrument to give 100% transmission of light. Use a blue or neutral filter.
4. Determine the amount of light transmitted by the 10^{-3} dilution.
5. To check that the absorption of light by the medium is not altered by the bacterial growth in it, it is necessary to centrifuge the bacteria down from the medium and then re-suspend

them in fresh medium of the same volume. The reading obtained on the colorimeter should be the same as before. If it is not, the amount of light being absorbed by the medium is being affected by the bacterial growth and it is therefore necessary to centrifuge the bacteria down from each dilution and to re-suspend them in the same volume of fresh medium.

A centrifuge operating at least 3,000 g, and preferably 4,000 g is required to separate the bacteria from their media.

6. Determine the amount of light transmitted by each of the dilutions in turn, centrifuging and re-suspending in fresh media if necessary.
7. After 2 days count the number of bacteria on the dilution plates. It is probably wise to check the plates after 1 day as some bacteria may grow rather quickly.
8. In plotting the graph of number of bacteria against the turbidity, express the latter as the optical density which is proportional to the cell concentration. This is a function of the negative log of the percentage transmission and therefore is expressed as $2 - \log$ of the colorimeter reading.
(Note that the EEL colorimeter scale is a log scale.)

Experiment

Use this calibration plot to determine the number of bacteria in a sample of unknown concentration. Follow the growth of this culture with the aid of the plot and the colorimeter. Readings should be taken as often as possible. Use your readings to plot a growth curve.

Requirements

Apparatus

A 250 ml flask containing 99 ml of sterile water

Six McCartney bottles, each containing 9 ml of sterile broth

Seven 1 ml pipettes, each one must be wrapped in brown paper and sterilised in an oven

Six sterile Petri dishes

Photo-electric colorimeter with a blue or neutral filter

Eight 10 ml colorimeter tubes

1 McCartney bottle containing 10 ml of sterile nutrient broth

Centrifuge capable of producing 3,000–4,000 g

GENETICS

The demonstration of mitosis in root tips

Procedure

1. Cut off a root tip about 1 cm in length.
2. Place it in watch glass of acetic orcein stain and NHCl.

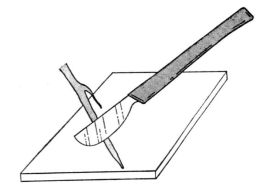

3. Warm the preparation over a spirit flame for about 5 minutes. The acid helps to macerate the cells so that the stain can penetrate more easily.

4. Place the root tip on a clean glass slide.
5. Cut off the tip 3 mm and discard the rest.

6. Add 2–3 drops of acetic orcein stain.

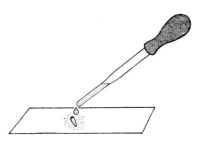

7. Break up the root tip with a needle, but keep the arrangement of cells on the slide so that you know which cells were part of the extreme tip.

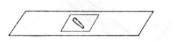

8. Place a coverslip over the preparation.

9. Cover with several layers of filter paper or blotting paper.
10. Squash with the thumb. Take care not to allow any lateral movement of the coverslip.

11. It may improve the preparation to warm it over the spirit lamp for a moment.
12. The preparation is now ready for examination.

Requirements

Apparatus

Spirit lamp
Watch glasses

Microscope, slides and coverslips
An oil immersion lens is useful

Reagents

Acetic orcein stain. For the preparation of this stain, see the 'meiosis in locusts' experiment (page 61)

Acetic orcein/NHCl mixture (10 parts of stain to one of acid)

Biological

Broad bean root tips are suitable for this experiment
Soak the seeds in water for 24–36 hours, until the testa has cracked at the hilum. Remove the testas to

ensure more even germination. Then plant the beans in vermiculite which must be kept moist. After about 48 hours when the radicle is 2 cm long, cut off the terminal 3 mm. This stimulates the growth of lateral roots when the seed is replanted. There should be a good crop in 5–7 days Onions are unsuitable in the autumn term because of dormancy

If metaphase figures are needed, soaking the root tips in 0·2% colchicine for 24 hours just before use, will inhibit spindle formation

The demonstration of meiosis in locusts

Procedure

1. Select fifth instar male *Schistocerca* hoppers that are almost ready to moult. *Locusta* hoppers can be used but the dissection is more difficult owing to the greater amount of fat.
   Anaesthetise in carbon dioxide or ether. This should be done by the teacher in the preparation room.
2. Kill the hopper by cutting off the head.
3. Remove the legs and wing buds.
4. Pin the hopper out in a dish covered with wax, dorsal side uppermost.
5. Cover with water.
6. Make a long incision along the mid-dorsal line.
7. Expose the body contents by turning back the body wall.
8. The testes, looking like a single, oval yellow body, lie above the gut between segments 5 and 6.
9. Remove the testes and remove as much of the fat body as possible.
10. Place a small piece of the testes on a clean, dry slide.
11. Place a second slide on top of this and gently squash them together.
12. When the slides are separated, there are two meiosis preparations ready for staining.
13. Add several drops of acetic orcein stain to the slide and place the coverslip in position.
14. Warm the slide over a spirit lamp or very low Bunsen flame for 10–15 seconds.
15. The preparation is now ready for examining under the microscope.
    This preparation should last for several hours.

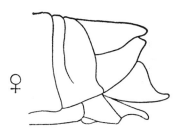

Schistocerca gregaria.
(Desert locust)

Locusta migratoria migratorioides. (Migratory locust)

Requirements

Apparatus

Dissecting dish covered with wax
Spirit lamp

Microscope, slides and coverslips
An oil immersion objective is useful

61

Reagents

Acetic orcein stain. This should be made up by boiling 45 ml of glacial acetic acid with 55 ml de-ionised water. Add 0·5 g of powdered orcein and boil gently, stirring from time to time, until the volume has been reduced to two-thirds of the original. When the solution has cooled, filter

Biological

Fifth instar male *Schistocerca* (Desert locust) hoppers

The preparation of a squash of 'Drosophila' salivary gland to show whole giant chromosomes

The salivary glands of *Drosophila* have developed very large cells which are controlled by proportionately enlarged nuclei. In the course of enlargement the chromosomes have duplicated themselves many times. Each duplicate remains lined up parallel to all the other duplicates of the same chromosome. It is possible for a single chromosome to consist of a thousand duplicates lined up together.

The banding which can be seen on these chromosomes represents small density differences along the chromosome which is magnified by the increase in number of the chromosomes. This banding is inherited as a constant feature. It can be used to indicate the position of a specific gene locus.

Note also the chromocentre which is common to all the chromosomes. This consists of the combined centromeres.

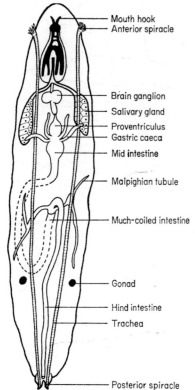

Mouth hook
Anterior spiracle
Brain ganglion
Salivary gland
Proventriculus
Gastric caeca
Mid intestine
Malpighian tubule
Much-coiled intestine
Gonad
Hind intestine
Trachea
Posterior spiracle

Procedure

To obtain larvae

When larvae are beginning to appear in the culture, slow down their development by keeping them at 18°C.

Add a small amount of yeast suspension each day.

The larvae are ready for dissection when a few have begun to pupate on the sides of the culture vessel.

Larvae should be chosen which are large and almost motionless. Larvae can be kept in this state in the refrigerator at 8°C for a week.

Dissection

1. Place the larva in a few drops of 'dissecting' solution on a slide.
2. Using a 20× binocular microscope and two dissecting needles, place one needle across the rear half of the larva to prevent movement.

3. Place the point of the other needle just behind the mouth parts.

4. Move the two needles apart gradually. This will draw the head and salivary glands away from the body.
   The glands can be recognised as two long sausage shaped objects with a fat body lying along one side. They should appear crystalline compared to the surrounding structures.

5. Dissect away any other tissues which may be attached to the glands. Brain ganglia will probably be attached.

6. Transfer the glands to the 'staining' solution on a slide which had been previously cleaned with a lens tissue.

7. After 10 minutes in the 'staining' solution, place a number 0 or number 1 coverslip over the preparation.

8. Apply pressure to the coverslip with your thumb through several layers of blotting paper. This should spread out the chromosomes.

9. Examine the preparation under the high power objective using a green filter. Oil immersion can be used but this is not essential.

Requirements

Apparatus

20 × binocular microscope
Ordinary microscope with oil immersion objective and green filter (oil immersion is not essential)
Microscope slides

No. 0 and no. 1 coverslips
Lens tissues
Dissecting needles
Blotting paper

Reagents

'Dissecting' solution. Either isotonic saline, i.e. 0·67% NaCl, or, a solution made up thus:
1 g of Gurr's natural orcein added to 45 ml of hot glacial acetic acid. After removing this from the heat add 30 ml of de-ionised water and 25 ml of 85% lactic acid. When filtered, this solution will keep for some time

'Staining' solution. This consists of 2 g of synthetic orcein dissolved in 50 ml of hot glacial acetic acid. After removing from the head add 50 ml of 85% lactic acid. After filtering, this solution can be stored
(If natural orcein is used throughout, the staining is somewhat lighter)

Biological

Drosophila larvae

An experiment to show the consequences of genetic segregation

The purpose of this experiment is to observe the segregation of genetical characters which are readily visible and to observe the relationship of crossing-over to segregation at each of the divisions of meiosis.

Monoploid ascospores ● ○

Monoploid mycelium

Monoploid mycelium

Hypha with Conidia

Mycelium

Mycelium

Immature Perithecia

Fertilisation

Diploid

First division of Meiosis

Second division of Meiosis

Mitotic division

Mature Perithecium with many asci, each containing 8 monoploid ascospores

These aims can only be achieved if an organism is used in which fertilisation does not follow meiosis immediately and in which the gametes are not the only monoploid cells.

In *Sordaria fimicola* it is possible to look directly at cells which are equivalent to the gametes of higher organisms and see the genotypes prior to fertilisation.

The details of the life cycle of *Sordaria* are shown in the diagram. The hyphae produce both conidial spores and sac-like bodies known as perithecia. When a conidium enters a perithecium, fertilisation occurs. The entering spore fuses with one of the many special cells in the perithecium. The diploid zygote grows to form an ascus.

Within the ascus the zygote nucleus goes through the two divisions of meiosis to give four monoploid nuclei. Each of these then divides mitotically. This produces eight ascospores.

Until the ascus containing the ascospores breaks up, the eight spores are arranged in a linear order, providing an accurate diagrammatic record of what has happened at meiosis. This is because all the products of one meiotic division are contained in one sac and cannot be confused with the products of other divisions, and also because the position of a particular ascospore in the sac can be referred to the actual position of a nucleus in meiosis, which is determined by the orientation of separating chromosomes on the spindle.

The eight ascospores then represent an ordered tetrad of duplicate meiotic products.

In this experiment a wild-type strain is mated with a mutant strain which has white ascospores. The wild-type has black ones.

Procedure

1. Using sterile technique, inoculate a sterile Petri dish containing cornmeal agar, with wild-type *Sordaria* culture.
2. Then inoculate the dish an inch away from the previous inoculation with the mutant culture.
3. Incubate the dish at 25°C.
4. Observe each day with binocular microscope until you observe perithecia. To the naked eye they will appear as tiny black specks.
5. Pick off the hybrid perithecia with a needle and crush them in a drop of water under a coverslip.

Results

Observe as many perithecia as possible; if necessary pool the results of all the members of the class.

Draw a diagram representing the arrangement of ascospores in the ascus each time a new arrangement is seen.

Count the number of perithecia with each type of arrangement.

Questions

1. For each of the ascospore arrangements, at which division of meiosis has segregation taken place?
2. Explain why the genetic segregation of alleles can occur at the second division of meiosis even when there is only one pair of segregating alleles.
3. What does the relative frequency of first and second division segregation depend upon?

Requirements

Apparatus

Sterile Petri dish   Binocular microscope
Inoculating loop   Mounted needle
Microscope slide and coverslip

Reagents

15 ml of cornmeal agar made up thus: Yeast extract 1 g
Difco cornmeal agar (from Oxoid   Make up to 1 litre with tap water
Ltd) 17 g

Biological

Sordaria fimicola strain C7h with black ascospores
Sordaria fimicola strain C7h(1) with white ascospores

Both of these strains can be obtained from the Department of Education and Science Laboratories, Ivy Farm, Knockholt, Sevenoaks, Kent and from Philip Harris Ltd

---

The use of 'Drosophila melanogaster' in the study of genetics

Life cycle

The egg (0·5 mm)

Adult females are capable of laying eggs two days after emerging from the pupal case. Laying increases for about a week until it is a maximum of 50–75 eggs a day. They are laid on the surface of the food.

The egg is ovoid with two thin stalks projecting from the end. These are flattened at their ends to act as 'water-wings', preventing the egg from sinking below the surface of a semi-liquid food medium.

Eggs can be seen with the naked eye on the surface of the food.

The larva (4·5 mm)

This is white, segmented and maggot-like.
Note the black mouth parts (jaw hooks).

It burrows in the food, feeding voraciously.

It undergoes two moults. After the final moult, it crawls out of the food on to a relatively dry surface to pupate.

The pupa (3 mm)

The cuticle becomes hard and pigmented.

The adult (2 mm)

In the newly-emerged fly the wings have not expanded and the body is light in colour. These change in a few hours.

Emergence mostly takes place during the early hours of the morning (7–11 a.m.), although it is diurnal.

The female matures in 12–18 hours and then lives for about 26 days. The male lives for about 33 days.

After copulation the females retain the sperm in two spermathecae and in the uterus. Therefore once a female has been mated it is not possible to use her again for genetic work.

Chronology of the life cycle: (at 25°C)		
Hours		*Stage*
0	—	Egg fertilised prior to laying
22	0	Egg hatches
47	25	First larval moult
70	48	Second larval moult
118	96	Pupation occurs
214	192	Emergence of adult

Note that egg laying may be delayed for two days after etherising.

The culture of Drosophila

The preparation of food

It must pour easily when hot and set firmly when cold, and remain firm when the adults emerge.

Soak 72 g of oatmeal in 120 ml of water and dissolve 35 g of black treacle in 40 ml of water. Boil 6 g of agar in 400 ml of water and add a pinch of Nipagin (methyl p-hydroxy-benzoate). The Nipagin should be made up by dissolving 0·1 g in a small volume of 95% alcohol. Add the oatmeal to the agar.

Boil all together for 15 minutes, stirring continually. If necessary add more water. This amount is sufficient for 12 third pint milk bottles or 60 3″ × 1″ specimen tubes. There should be about an inch of food in each bottle.

Pour the food when it is sufficiently mobile to pour, but will set on cooling. To keep the sides of the bottles clean, it is best to pour the food through a filter funnel, using a glass rod flattened at one end as a ram-rod. Use a plastic funnel with a wide neck.

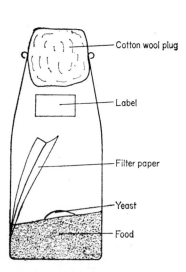

Cotton wool plug

Label

Filter paper

Yeast

Food

When the food has set add a few drops of a yeast suspension.

A folded piece of paper towelling (non-medicated) or filter paper should be placed between the glass and the food on one side.

After the steam has escaped, seal the bottle with a cotton wool plug. **This must be tight to prevent stray flies entering the bottle.**

In the hot weather the food often liquefies rather too readily. This can partly be overcome by using 'quickgel' (sold by most grocers) instead of the agar.

Sterilisation of materials

Nipagin is an anti-mould. If this is not available, it can be replaced by 0·2% propionic acid. Once the yeast culture is active, the danger of contamination is considerably reduced and therefore flies should be introduced into the bottle as soon as possible, as these spread the yeast across the food surface and so help it become established.

Most text books recommend that before the food is added, the cotton wool stoppered bottles should be autoclaved at 121°C for 15 minutes. In practice if the bottles are thoroughly washed out and the flies are added soon after the food is added, the yeast culture seems to be able to compete with any form of contamination. Nevertheless it is wise to periodically autoclave the bottles.

Temperature

The usual temperature for experimental matings is 25°C but they can be kept at room temperature. This will slow the life cycle down but this can be overcome by grouping them around a bench lamp if an incubator is not available.

Remember that temperatures of 28–30°C will make the progeny sterile. Prolonged cold (10°C) will kill the flies.

Mites and moulds

Drosophila often becomes parasitised by mites and moulds. They can be controlled by repeatedly subculturing or by keeping the flies strictly at 25°C in an incubator. The life cycle of the mite is longer than that of the fly.

The mites can be recognised as tiny pinkish specks on the glass above the food and on the flies.

Mite-infested bottles should be washed in a strong detergent.

The most common mould is *Penicillium*. This dries the food and so starves the flies. It is possible to use cultures which are contaminated if more yeast is added to compete with the other fungi.

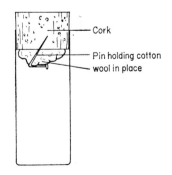

Cork

Pin holding cotton wool in place

Techniques in handling the flies

Etherising

This is necessary to sex or score the results of experimental matings. They are **lightly anaesthetised** with ether.

A 3″ × 1″ specimen tube is suitable if it is stoppered with cork to which a piece of cotton wool is pinned. **The neck of this tube must fit the neck of the culture bottle tightly.**

The cotton wool should be soaked in ether and the tube stoppered for a few moments to allow the ether to saturate the atmosphere of the tube. **Avoid having liquid ether in the bottom of the tube when adding the flies.**

To transfer the flies to the etheriser:

1. Knock the culture bottle sharply on the palm of the hand to release the hold of the flies and to knock them to the bottom of the bottle. If some flies still hold on, twirling the bottle around usually succeeds.
2. Quickly remove the stopper from the ether bottle and the plug from the culture bottle and hold the two necks together.
   Ether is heavier than air, **therefore the etheriser should be on the bottom.**
3. Gently tap the culture bottle so that the flies fall into the etheriser.
   Remember that Drosophila are phototropic and so if any difficulty is experienced, the etheriser can be held towards a light, but avoid getting ether fumes in the culture bottle.
   If the food is very soft, make use of the phototropic response, adding the ether soaked stopper after adding the flies. In this case it is very easy to allow liquid ether to run into the etheriser.

When the flies stop moving, pause for 10–20 seconds and then shake them out on to a tile.

Recognise over-etherised flies from:

1. Wings extended straight above body.
2. Tip of abdomen is curled under.
3. Legs are bunched up under the body.

Usually they remain etherised for 5–10 minutes, but if it is necessary to re-etherise them this can be done on the tile by holding a watch glass or Petri dish lined with gauze, lint or merely paper towelling over the flies. The lining should be soaked with ether.

Sexing

To distinguish the sexes note the following:

1. Female is longer than the male.
2. Tip of the female's abdomen is more pointed than that of the male.
3. When viewed from the dorsal surface the male abdomen tip is more heavily pigmented.

With practice the above features can easily be seen without the aid of a lens.

With a binocular microscope (×20) it is possible to see

1. External genitalia.
2. The sex comb on the foreleg of the male.

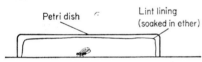

Petri dish   Lint lining (soaked in ether)

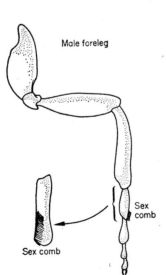

Male foreleg

Sex comb

Sex comb

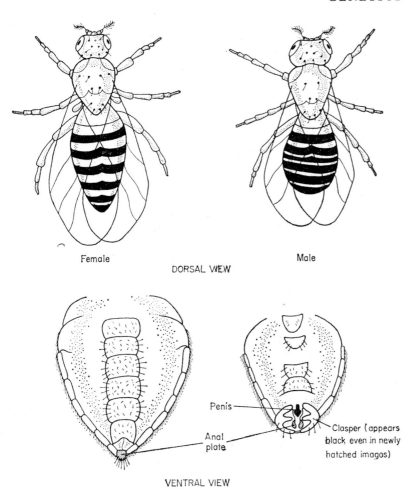

Female   Male

DORSAL VIEW

Penis

Anal plate

Clasper (appears black even in newly hatched imagos)

VENTRAL VIEW

As a fly is sexed it should be moved to the left of the tile if it is a female and to the right if it is a male.

A camel hair brush or a seeker are suitable tools for moving the flies.

Isolating virgins

As females can store sperm, it is essential that **only virgins are used for experimental breeding work.** A female matures when it is 12–18 hours old.

A virgin can be recognised from its ash-grey colour and the presence of pupal faeces, a black spot seen through the wall of the abdomen.

To collect virgins

1. Shake out all the flies from the culture bottle.
2. No more than 12 hours later, shake out the newly hatched flies. Any females here will be virgins.

In practice errors are negligible for school purposes if the cultures

69

are cleared just before morning school and after afternoon school. If sufficient flies are available, only use those collected in the afternoon.

The flies should be placed in bottles according to sex and mutant type until required for mating purposes.

Males may be collected at any age and time.

Setting up matings

The initial crossing should include 5–6 males and 5–6 females. A reciprocal cross should always be set up, if not by the same person, by another member of the set.

After 2 days the parents should be transferred to a fresh bottle to prevent overcrowding when the offspring begin to emerge.

When placing etherised flies in a culture bottle, do not allow them to come in to contact with the food until they have fully recovered. Lie the bottles on their sides.

A bent index card is a convenient means of transferring the flies from the tile to the culture bottles.

When the flies have recovered stand the bottles upright.

As soon as the mating is set up the bottle should be **labelled with the nature of the cross and the date.** The mutant symbol of the female is always written first and so there is no need to indicate which is the female.

Third or half pint milk bottles are usually recommended as stock culture bottles and $3'' \times 1''$ specimen tubes for experimental matings. In practice we have found that milk bottles are also the best for experimental matings; overcrowding is reduced and hence the number of offspring is increased quite considerably without the need for repeated transfer of the parents to new bottles (for the results to be statistically valid as many offspring as possible should be obtained). Secondly moulds also appear to be less common in larger bottles. This may be because there is a greater variation in temperature in the small tubes when kept at room temperature and this may favour mould growth; also small tubes dry out much more rapidly. Considering the time involved in isolating virgins, it seems to be well worth while to use third pint milk bottles.

Scoring

Arrange the flies in a single line and separate on one of the characters segregating, making two rows. Divide each row into males and females. The scores should be taken regularly and the counted flies removed from the bottle.

Discarding flies

Unwanted flies must not be allowed to fly away, as mutant flies loose in the laboratory are a nuisance. A wide mouthed specimen jar containing 70% alcohol should be provided as a 'morgue'.

Timetable for classwork

A possible timetable is as follows:

Day 1. Set up the mating.

Day 3. Transfer parents to a second culture bottle.

Day 4. Kill the parents unless they are required for back-cross experiments. In which case they should be separated into separate bottles.

Day 13. Adults should begin emerging. **Score all the flies that emerge. Do not give up after a couple of days as some of the mutant types commonly used are associated with different developmental rates.**

Many books give a timetable of this type but because temperature variations make it difficult to keep to this timetable, **it is best for the student to work out his own timetable.** In fact it is best if the bottles can be examined each schoolday. This should only take a few moments most days; sufficient to determine the stage of the culture.

Requirements

For preparing the food and stock cultures.

Apparatus

Gas or electric ring

Filter funnel for pouring the food into the bottles

⅓ pint milk bottles

Paper towels or filter paper

Cotton wool

Biological

Yeast, 1 oz
packets of dried yeast are suitable

Reagents

Oatmeal (porridge)

Black treacle

Agar (quickgel is useful in the summer)

Nipagin (methyl p-hydroxy-benzoate). This can be obtained from British Drug Houses

For the students' experimental work

Apparatus

⅓ pint milk bottles of food medium or 3″ × 1″ specimen tubes of food medium

Tile

Camel hair brush or seeker

Index card, bent for use in transferring flies from the tile to the bottles

Watch glass or Petri dish lined with

absorbent material to serve as a re-etheriser

3″ × 1″ specimen tube and cork to fit it to serve as an etheriser

Hand lens and if possible a binocular microscope, × 10

Labels or chinagraph pencils

Reagents

Ether

Biological

Wild type and several mutant types of *Drosophila*

71

A cross between two types of 'Drosophila' differing from each other in only one visible way

The purpose of this experiment is to determine if it is possible to deduce any laws governing the inheritance of the characters in which these flies differ.

Procedure

1. Mate a female wild-type fly with an ebony-bodied mutant fly. Set up the reciprocal cross.
   Alternatively, a wild-type fly can be mated with either a brown-eyed fly, a vestigial-winged fly or a scarlet-eyed fly. These mutants are either more difficult to culture or more difficult to score for the beginner.
2. The parents must be transferred to other bottles after 2–3 days. If it is not desired to allow the parents to continue breeding they should be placed in separate bottles or tubes and kept for later use.
3. When the F_1 flies (i.e. the first generation of offspring) emerge, observe the nature of the character being studied and separate 5–6 of each sex into a fresh mating bottle to produce the next generation. **It is not necessary to use virgins here.**
4. Concurrent with this mate 5–6 of this F_1 generation with flies from the parental mutant type.
5. After 2–3 days, clear the parents from these newly set up matings.
6. When the offspring begin to emerge, i.e. the F_2 and the progeny of mating the F_1 and the mutant parent, score the results as outlined on page 70 and dispose of the flies.

Results

These could be set out as follows:

FEMALE PARENT × MALE PARENT		ORIGIN OF FEMALE		ORIGIN OF MALE	
. . . .		. . . .		. . . .	
DATE OF MATING		DATE OF FIRST LARVAE			
DATE OF FIRST PUPAE		DATE OF FIRST ADULTS			

DATES OF COUNTING	PHENOTYPE 1		PHENOTYPE 2		PHENOTYPE 3	
	female	male	female	male	female	male
TOTALS						

Questions

1. Compare your F_1 results with other students performing the same mating. Can you draw any conclusions from this result?
2. Can the numbers of the different types of flies obtained in the F_2 generation be made to approximately fit a simple ratio?
3. Can they be made more easily to fit a simple ratio if the results of all the students performing the same mating are pooled?
4. Why did the mutant character remain concealed in the F_1, but then reappear in the F_2?
5. Are we wrong in simplifying our ratio, or can we consider the experimental ratio to differ from the simple one by chance alone?

To analyse these results, make use of the **Chi-squared method** (page 298).

6. Similarly compare the result of your mating the F_1 with the parental mutant (i.e. back-cross or test-cross).
7. What does this test-cross mating demonstrate?
8. What can you say about the likelihood of fertilisation of an egg cell by a gamete carrying a particular type of character-determining particle?
9. If the most abundant phenotype in the F_2 generation is inbred, what would you expect the F_3 to be like?

Requirements

Apparatus and Reagents

As given in general techniques section

Biological

Wild-type flies of both sexes
Mutants of both sexes: ebony body, the most suitable for ease of scoring and culturing
or, brown eye, easy to culture and fairly easy to score
or, vestigial wing, easy to score but not always easy to culture unless an incubator is being used. If food is slightly soft, they become trapped on it. In an experiment in which every fly must be counted, this is disastrous

Setting matings up in advance

At sixth form level this should be unnecessary as the student should have both the time and the interest to do his own work, particularly as after the initial periods spent teaching the basic technique, much of the work has to be done out of normal period times. If the bottles of culture medium are provided, it should only take each student a few minutes each day at the most. For staff to do this for each student, a generous allowance of technician time will be required. If setting up in advance of classwork, for scoring F_1, set up 14 days before required; and for F_2, 28 days before required. If the temperature is rather low allow more time.

Biochemical evidence for the particulate theory of inheritance

The red eye pigment of a fruitfly is made up of several compounds which fluoresce under short wave ultraviolet light.

These compounds are present to varying extents in the mutants. The amount present also depends on whether the mutant gene is present in the heterozygous or homozygous form. The different amounts and different kinds of compounds are the expression of the metabolism of the fly. Changes in metabolism will lead to changes in these compounds, and changes in metabolism are caused by genes.

In this experiment pure red-eyed and pure white-eyed flies are being mated to produce red-eyed F_1 flies. The chemical nature of pigments in the parents and the F_1 flies is then compared by means of paper chromatography.

Procedure

Preparation of fly pigments. (For details of chromatographic techniques see page 110)

1. Use three female flies, one parental red-eyed fly, one parental white-eyed fly, and one red-eyed F_1 fly. (This is the result of a mating between female red-eyed fly and a male white-eyed fly.)
2. Immerse each of the flies in a separate, clearly-labelled test tube of alcohol for 1 minute.
3. Then pour off the alcohol and replace it with de-ionised water. Boil the flies in the water for 1 minute.
   This coagulates the proteins and therefore improves the resolution of the pigments.
4. Prepare the filter paper in the usual way. Rest it on a tile to prevent loss of pigment through the paper when squashing.
5. With a clean razor blade cut off the flies' heads and place them on the appropriately labelled origins.
6. Crush them on to the paper with a clean spatula.

Solvent

n-butanol. 60 parts by volume.
glacial acetic acid. 15 parts by volume.
water. 25 parts by volume.
If a 2-way separation is being tackled, the following solvent could be used:

n-propanol   66 parts by volume
water   32 parts by volume
0·880 ammonia   2 parts by volume

Running time

14–16 hours.

Locating agent

Ultraviolet lamp.
Examine under both long and short wave ultraviolet light if possible as some pigments are more easily seen under one than the other.
WARNING. Short wave ultraviolet light, i.e. less than 3100Å units, is dangerous to the eyes if viewed directly. This will cause extreme pain. The effect on the retina is the same as the effect of extreme sunlight on the skin. You must particularly realise that the pain will not appear until some time after contact with the ultraviolet lamp.

Results

Indicate the outline of each colour with a pencil, and indicate the nature of the colour.
This must be done soon after the paper dries as these pigments are not stable in light for very long.

Questions

1. How does this experiment provide evidence for the particulate nature of inheritance?
2. Can you find a difference in the pigments present in the parent with red eyes and the F_1 flies?
3. If you are able to recognise a difference between these flies, what conclusions would you come to about the control mechanism of metabolism?

Requirements

(in addition to those needed for all chromatography experiments)

Apparatus

Tile
Razor blade
Spatula
3 test tubes
Ultraviolet lamp

Reagents

Alcohol
n-butanol
glacial acetic acid
n-propanol
0·880 ammonia

Biological

Wild-type *Drosophila*
White-eyed mutant *Drosophila*
The F_1 produced from a cross between the female red-eyed fly and the male white-eyed fly

A cross between the wild-type 'Drosophila' and the white-eyed mutant

The purpose of this experiment is to discover if this mating obeys the same laws as deduced in the previous experiment.

Procedure

1. The mating should be carried out as in the previous experiment. A reciprocal mating **must** be set up.
2. There is no need to spend time on the test-cross.
3. Observe the nature of the F_1 flies and if there are several types, score as for the F_2 generation in the last experiment. Pay particular attention to the sexes of all the flies in the F_1 generation.
4. Set up the mating of these F_1 flies for the production of F_2 flies.
5. When the F_2 flies begin to emerge, score and record as in the previous experiment.

Results

If this mating is obeying the same law as was deduced in the last experiment, we could say that these results should produce the same simple ratio.

To decide if the deviation from this is due to chance alone use the **chi-squared method** thus:

$$\chi^2 = \Sigma \left[\frac{(a - b)^2}{b} \right]$$ where a represents the observed score, b the

expected score and Σ 'the sum'.

If there are 2 classes (i.e. 2 phenotypes in this case), there will be one degree of freedom.

In the chi-squared tables read across the one degree of freedom line to the value of chi-squared approximating most closely to the value you have calculated. From this figure read off the probability of this value being obtained.

Within the limits of biological variability, it is usual to accept 5% as a significant probability. If a probability of less than this is obtained, we usually consider that it could not have happened by chance alone and therefore we must look for some other mechanism to explain this result.

N.B. **It is important to treat each of the reciprocal experiments quite separately.**

Questions

1. Try to offer an explanation for any differences you obtain in the results of your mating with that of its reciprocal cross.
2. Does this mating obey the same law as the mating performed in the last experiment? If you find it difficult to make it fit the law

can you suggest how to modify the law to fit the results you have obtained?

Requirements

Biological

Wild and white-eyed flies of both sexes

A cross between two 'Drosophila' differing from each other in two visible ways

The purpose of this experiment is to discover if the factor controlling one of these characters can in any way influence the occurrence of the factor controlling the other character.

Procedure

1. Mate the vestigial-winged mutant with the ebony-bodied mutant.
2. Set up the reciprocal cross.
3. Continue as in the previous experiments.

Questions

1. What is unusual about the F_1 generation as compared with this generation in the previous experiments?
2. Can you make the products of the F_2 generation fit a simple ratio?
3. What conclusion would you draw from these results?

Requirements
Biological
Vestigial-winged and ebony-bodied mutant flies of both sexes

A demonstration of multiple alleles

Multiple alleles consist of a series of genes from independent mutations at the same locus on a chromosome but not more than two of the genes can occur together in a normal diploid fly.

Compare this with the cases met so far where only two alleles exist at any one locus.

The locus being studied here is that of the *w* locus on the *X* chromosome.

You have already discovered that the wild-type, red-eyed condition is dominant to the white-eyed condition. In this same series of alleles there is an apricot-eyed mutant allele.

In this experiment, you will discover the position of this gene in the series, and prove that the three genes must be allelomorphic.

Procedure

1. Mate female apricot-eyed flies with male white-eyed flies.
2. Continue as in the previous experiments, observing the eye colours of the F_1 and F_2 generations. Pay particular attention to the numbers of each sex within each phenotypic group.

Results

Set these out as in the previous experiments.

Questions

1. How many phenotypes are there in the F_1 generation?
2. Do these phenotypes bear any correlation to sex?
3. If so, can you explain this correlation?
4. How does this experiment prove that these are multiple alleles of one locus and not alleles from different loci?
   (The alleles being considered are those for red eye, apricot eye and white eye.)
5. If these genes were not allelomorphic, what would be the eye colour of the F_1 flies in the mating between apricot- and white-eyed flies?

Requirements
Biological

Female apricot-eyed flies and male white-eyed flies

Evidence for the chromosome theory of inheritance

Genetic factors are inherited on chromosomes and if they are on different chromosomes they will be inherited independently of each other. This has already been seen when making a dihybrid cross. This is the basis of Mendel's second law.

There may however be cases where the assortment of two or more genetic characters is not independent. Each gene may not have an equal chance of occurring independently of the other. These genes may seem to be 'linked'. This at first sight may appear to be an exception to Mendel's law but it can be explained by the 'chromosome theory'.

In the male *Drosophila* complete linkage within a gene is always the case but in the female there may be an exchange of corresponding segments between the chromatids carrying the alleles of homologous chromosomes. This gives rise to offspring carrying new combinations of characters not previously seen in the parents. This is termed **crossing-over.**

In this experiment you will confirm the presence of genes on chromosomes by comparing the offspring from two genes 'in coupling' and 'in repulsion', and you will confirm the existence of crossing-over.

Genes **in coupling** are two recessive genes on one chromosome while their dominant alleles are on the other chromosomes of a pair.

Genes **in repulsion** are one recessive and one dominant on one chromosome and the alternative dominant and recessive alleles are on the other chromosome of the pair.

Procedure

Part I. (a) **In coupling**
1. Mate female white eye, miniature-winged flies with male wild-type flies.
2. Score the F_1 generation when it emerges and mate these flies to produce the F_2 generation.
3. Score the F_2 generation when it emerges (use the same phenotypes in part (*b*)).

Part I. (b) **In repulsion**
1. Mate female white-eyed, normal-winged flies with male red-eyed, miniature-winged flies (use the recombinant phenotypes from the F_2 generation of part (*a*)).
2. Continue as for part (*a*).

Results
Calculate the number of each phenotype as a percentage of the total number of flies in the F_2 generation.

Classify each of the phenotypes as 'parental' or 'recombinant', i.e. as similar to one of the parental stocks or as a new combination. If the genes were assorting independently the recombinant phenotypes would be equal in frequency to the parental phenotypes.

Use the chi-squared method to test this hypothesis.

Questions
1. In each part of the experiment how many phenotypes would you expect in the F_2 generation in the light of your experience with the previous experiments in this book?

2. Can you explain how the actual, observed number of phenotypes comes about? Try to produce an answer without discarding Mendel's laws.
3. Does the observed ratio of phenotypes in the F_2 generation fit the expected? Can any deviation from the expected be by chance alone?
4. If not can you suggest any way of explaining these results without destroying the apparent universal applicability of Mendel's laws?
5. How does this experiment show that inheritance is carried on by discrete particles known as genes on chromosomes?

Procedure

Part II

1. Mate female wild-type flies with black-bodied (do not confuse with ebony), vestigial-winged flies.
2. When the F_1 generation appears back-cross virgin F_1 females to the parental mutant stock.
3. Score the F_2 generation when it emerges.

Results

Treat the results as for those in the earlier part of the experiment.

Questions

1. Can you account for the number of phenotypes you obtained?

Requirements

Apparatus and Reagents

As for all *Drosophila* work

Biological

Wild-type flies of both sexes
White eye, miniature wing mutants
of female sex

Black bodied, vestigial-winged flies
of male sex

The loci of some 'Drosophila' genes (a linkage map)

When the gene is that which produces the wild-type character at any of these loci it is usually represented thus:

y^+, w^+, or vg^+. In the heterozygous state this will be w^+w or vg^+vg.

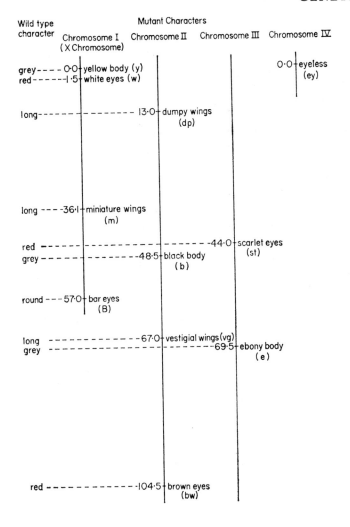

An exercise in chromosome mapping

In *Drosophila*, the genes for white eyes, yellow body and miniature wings are all sex linked. Appropriate matings reveal that the white and yellow genes consistently show the same frequency of crossing over. Miniature and yellow genes show a different but consistent value for crossing over also.

It would appear that crossing over between pairs of linked genes occurs at quite characteristic and stable frequencies, but these frequencies may differ widely, depending upon the pairs involved.

This generalisation is consistent with the idea that each gene has a definite locus in its chromosome.

This conclusion is supported by the fact that cross-over values can be used to map 'distance' relations between linked genes.

It is not possible to do this in terms of standard linear measurements, but chromosomes can be mapped in terms of percentages of cross-over obtained from genetic experiments. In this system a 5% crossing over between two genes, means they are situated 5 units apart on the chromosome map.

In order to illustrate the type of reasoning used, imagine there is a 6% cross-over value between genes X and Y, and that there is a 3% cross-over value between gene X and a third gene Z.

This means that we could map this part of the chromosome as in the figure. The gene Z could be mapped in either of these positions from the information so far available.

We can decide the position of Z by determining the cross-over value between the genes Z and Y. If we find this to be 3, Z must lie between the other two genes. If it is 9 it must lie to the left of X.

Procedure

In this exercise the positions of the sex linked genes for yellow body, white eye and miniature wing will be used. Three matings must be performed.

Mating 1

Use the data obtained in the previous exercise in which white-eyed, miniature-winged females were mated with wild-type male flies. Remember that the number of recombinant types is the cross-over value and this represents the distance between these two genes.

Mating 2

1. Cross a white-eyed, yellow-bodied female with a wild-type male.
2. Inbreed the F_1 generation.
3. Score the F_2 generation, noting the number of parental types.

Mating 3

1. Cross yellow-bodied, miniature-winged females with wild-type males.
2. Inbreed the F_1 generation.
3. Score the F_2 generation noting the number of parental types.

Results

From each of these matings you should have determined a cross-over value, i.e. the percentage of recombinant types.

Using this information and the method given in the theoretical account map the position of these genes on the sex chromosome.

Requirements

Apparatus and Reagents

As for all *Drosophila* work

Biological

White-eyed, miniature-winged female flies

White-eyed, yellow-bodied female flies

Yellow-bodied, miniature-winged female flies

Male wild-type flies

The inheritance of a quantitative character in 'Drosophila'

Mendel chose as parents for his crosses, individuals that differed by sharply contrasting alternative characteristics. Many of the characteristics of living things however cannot be described as 'sharply contrasting alternative characteristics'. These differences are often merely differences of degree on a continuous scale of measurement. These are **quantitative** characters. Often they seem to be subject to considerable modification by the environment.

Quantitative characters include such characters as height and intelligence in man, corn ear length, milk production by dairy cattle, but not height in peas. They do not include all characters involving measurement.

Quantitative characters, if sufficient readings are taken, form a normal distribution curve.

Procedure

1. Take 10–20 wild-type flies, taking an equal number of each sex, and count the number of bristles on the fifth abdominal segment. (If time permits, also count the number of the fourth abdominal segments as this extra information can be used in another exercise.)
2. Set the results out thus:

Segment	Fly number						Total
	1	2	3	4	5	6.....................	
4							
5							
Total							

3. Calculate the mean value for the number of bristles on the fifth segment.

4. Check that this mean value is representative of that of the total population.

To do this calculate the **standard error.** To determine the standard error the standard deviation must first be calculated. The standard deviation measures the variation around the mean, and it is calculated from the following formula,

$$\sigma = \sqrt{\left[\frac{\Sigma(fd^2)}{N-1}\right]}$$

where f represents the frequency of a class, d represents the deviation of a class, and N represents the number of classes.

The standard error, S.E. $= \dfrac{\sigma}{\sqrt{N}}$

It is customary to express any sample in terms of its mean $\pm$ its standard error. This will give an indication of its reliability as a representation of the number of bristles on the fifth abdominal segment.

Note that the standard error decreases the larger the sample you take.

You can check that your sample is representative of the total population by noting the mean values of the other students in your class. Their determination of the mean should almost always give a figure which lies within the range

$$\bar{x} + 2 \text{ S.E. and } \bar{x} - 2 \text{ S.E.}$$

and generally within the narrower range

$$\bar{x} + \text{S.E. and } \bar{x} - \text{S.E.}$$

If they do not, your sample cannot be representative of the population.

5. Mate 4–5 females with a low number of bristles on their fifth abdominal segment, with 4–5 male which have the higher numbers of bristles on their fifth abdominal segments.

6. Score the F_1 generation for the number of bristles on the fifth abdominal segment, and again determine the standard deviation and standard error for the sample.

7. Allow 4–5 flies of each sex to produce the F_2.

8. Score the F_2 as before.

Results

The results are best set out as histograms on graph paper as in the figure. On this chart the parents should form two quite distinct groups, while the spread and any segregation in the F_2 should readily be apparent.

Obviously the more flies that are scored in the F_1 and F_2 generations, the more reliable will be the results.

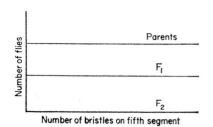

Compare the means and standard deviations of the F_1 and F_2 generations.

Questions

1. When sampling to obtain the parents for the experiment, why is it important that the sample is representative of the total population?
2. From your determinations of the mean, and standard deviations of the three generations, what conclusions would you draw about the inheritance of quantitative characters?

Requirements

Apparatus and Reagents

Binocular microscope ($\times 20$) or low power of monocular microscope

All usual materials for work with *Drosophila*

Biological

Wild-type flies of both sexes

The determination of the frequency of a gene in a population

So far we have been concerned with laboratory exercises, but if we consider a naturally occurring population, we find that some of the rules we have discovered may not commonly prevail.

There are two genetical differences between these populations.

1. A laboratory experiment is usually designed such that the relative frequency of alleles at a particular locus is fixed at a convenient ratio.

   Usually two homozygous parents are mated to start the experiment. **Therefore the alleles are introduced in equal frequency. In a naturally occurring population the relative frequency of alleles may vary considerably.** For there may be a tendency for one allele at a locus to be almost always homozygous.
2. The frequency with which different phenotypes and genotypes mate may vary in a natural population. Similarly the ability to produce offspring may vary. In the laboratory exercise the mating pattern is clearly laid down and it is impossible for mating to occur outside this system. This means that if there is a tendency for two flies not to mate, they have the choice of either not mating or of mating with this undesirable dam or sire.

With these differences in mind it is therefore necessary to be able to measure the frequency of a gene in a population. It is important that medical geneticists should be able to detect any increases in the frequency of harmful genes such as those causing

D

diabetes, mental illness of certain types, and muscular dystrophy. Today we must be able to assess the genetic effects of radioactive fall-out.

Do not assume that Mendel's laws bear any direct relationship to the frequencies of alleles in a population. Mendel's work applies to the patterns of inheritance in a family, given the dominance and recessiveness, the homozygosity or heterozygosity of the parents.

In this exercise we will consider a human gene, that which gives the ability to roll the tongue into a U-shape.

Procedure

1. We must resort to sampling. It must be as random a sample as possible to be representative of the population from which it is drawn. It must be fairly large, because if it is too small, chance may make it unrepresentative.

   In a truly random sample, members of the same family should not be included any more frequently than if they had been chosen by chance from the population.

   Therefore make the rule that **only one member of each family should be included in the sample.**

2. To avoid embarrassment to boys or inconvenience to staff, we have found the following system of sampling to be the most suitable:

   (a) Each student in the set should obtain data from one form in the school (no two students should use the same form).

   (b) The student should introduce himself to the form and member of staff the day before the sampling is to be done. On this day he should explain what he wishes to know, i.e. 'Can they curl their tongues into a U-shape?' The boys in the school will then have time in which to decide if they can do this or not.

   (c) The following day (a convenient time for these meetings to take place should be morning registration as only the briefest time is required each day) the student should ask the form to try to curl their tongues while he takes the score. It is important that **he actually observes whether they can curl or not.** The report of the boys may not be reliable, especially with younger boys who may not fully understand what is required of them.

3. Each student should also obtain data from his own family and then use this information to decide whether the ability to curl the tongue is a dominant or recessive trait.

Results

From your family data you will have discovered which allele is the dominant one. Represent this allele by A, and the recessive by a. Let us call the proportion of the population with this dominant gene p.

86

Let us call the proportion of the population with the recessive gene q.

Now if we add $A + a$, it must equal 1 or 100% of the population. Therefore $p + q = 1$.

Now if we know the frequency of the recessive gene q, $p = 1 - q$.

To find the frequency of one allele, we must turn to the concept of **gene pool.**

As we know of no reason why the genes should be distributed differently in the two sexes, we will assume that their frequencies are the same in both sexes.

Of all the sperm cells in the population gene pool, p of them will carry A allele and q will carry a allele.

Similarly for the egg cells.

To summarise,

	frequency of A	**frequency of a**
In sperm	p	q
In eggs	p	q

We can now think of these p and q values as probabilities. If a gamete were drawn from the gene pool at random, the probability of getting an A allele is p.

What proportion of the next generation will be AA, Aa, and aa? (Remember that the allele carried by the sperm is independent of that carried by the egg.)

Your answers to these questions produce the **Hardy–Weinberg equation.**

Using this equation, determine the value of q.

Then substitute in the equation $p + q = 1$, for p.

Find the frequency of the three genotypes.

Questions

1. What is the conclusion of the Hardy-Weinberg principle?
2. Can you prove that the frequency of the A alleles will be the same in the next generation as in the present one?

'Drosophila' genotype competition in a population cage

Procedure

Cages

A polythene sandwich box with a tightly fitting lid makes an ideal cage. Cut a small square out of the lid and stick very fine mesh wire over this to ventilate the cage.

Food can be supplied in 3 in × 1 in tubes cut in the base of the cage. A 9 inch long box could have 8 food tubes in its base. Usually these tubes are bought with corks; when the tubes are not in place the corks can be used to plug the holes in the box. Two holes in the sides can be used for transferring flies to the cage from clean tubes.

Food

The food tubes should be filled to the top with *Drosophila* medium. It must be firm as half should then be removed. With a suitable knife cut vertically through the food to the bottom of the tube. Discard half of the food plug.

Pour a few drops of yeast suspension on to the remaining half of the food plug.

Place a folded piece of paper towelling into the empty half of the tube. This should reach the bottom of the tube but it should not go beyond the top of the tube. This will permit gases to escape and provide a suitable site for pupation.

Fresh food tubes should be placed in the cage twice a week, but only one tube at a time. There should be only two present at the start.

Changing population cages

If the experiment is proceeding for some time this may be necessary. To do this first place all the food bottles in the new cage. The old cage will now only contain adult flies and dead ones. To transfer these flies to the new cage, seal a third pint milk bottle over one of the holes, cover the cage to exclude light and hold with the bottle towards the light. Tap the cage repeatedly. When a fairly large number of flies have entered the bottle, replace it with a new bottle. Lightly etherise the flies in the first bottle and transfer them to a clean empty food tube. Insert this into the cage.

If the population is very large it is not necessary to remove every fly from the old cage.

Setting up the mating

Lightly etherise the flies to be used, count them and transfer them to clean empty food tubes.

Avoid shaking the cage.

Ideally it should be kept in a constant temperature incubator but, unless the effect of temperature is being studied, this is not absolutely necessary.

Mark the food tubes with the date on which they were placed in the cage so that you know which one you are to remove each time.

Any stocks of *Drosophila* can be used. For instance the rate of disappearance of the white eye mutant from a culture of this and wild-type flies can be studied.

Observations

1. Estimation of selection in a population cage must be estimated against the **generation time**, rather than merely days. One way of doing this is to count the day on which the first food tube is placed in the cage as day one (assuming that eggs are laid on this day), and then observe until the first pupae appear. Mark the position of these pupae. Then periodically note whether they have become lighter in colour, signifying the emergence of the adults. This will be the final day of the generation time.

2. The frequency of the competing types must be estimated while the experiment is proceeding.

   To do this remove a slant of food with a knife from a food tube. Pour a drop of yeast suspension on to the remaining food in the tube and place it in the side hole of the cage for 24 hours.

   During this time the flies will have laid eggs on this fresh layer of food. Slice this off on to a microscope slide and count 125 eggs.

   Remove the remainder of the food from the slide and place the slide containing the known amount of eggs in a fresh culture bottle, containing food, yeast suspension and paper towelling. After a few days add a few drops of fresh yeast each day to provide optimum conditions for growth.

   When the adults emerge, score regularly until you are sure that all the flies are accounted for or that some of the eggs have perished.

Results

Estimating gene frequencies

If autosomal recessive mutant genes are used in competition with their wild-type alleles, the frequency of the mutant gene in the population can be estimated from the frequency of homozygous recessives, assuming that the Hardy–Weinberg equilibrium applies

$p^2 + 2pq + q^2 = 1$ where p is the probability of the dominant allele occurring and q is the probability of the recessive allele occurring.

Therefore the frequency of the recessive allele will be the square root of the frequency of the homozygous recessive phenotype.

The gene frequency of the dominant allele is then $1 - q$.

For sex linked genes, the gene frequencies in the male and females should be calculated separately. As the males carry only a single X chromosome, the frequency of the sex linked gene amongst the males is simply the observed frequency of the mutant among the males. Females have two X chromosomes and so the frequency of the gene is calculated as for normal autosomal genes.

Requirements

Polythene sandwich box with 8 holes in the bottom, each one large enough to take a 3″ × 1″ tube. These must make a tight fit with the box. There should be 2 similar holes in the sides. There should be stoppers to fit these holes when the tubes are not in place. There should be a hole in the lid covered with perforated zinc.

The other apparatus required is as for the other experiments with *Drosophila*.

The isolation of a mutant

Several factors can cause the production of mutants, but mutants resistant to ultraviolet light are possibly the most readily detected. If the mutant, because of the change it has undergone, is well suited to the new environment in which it has formed, it may become dominant over its parental strain in this new environment.

Spontaneously formed mutants resistant to ultraviolet light can be detected because only they can grow after treatment with ultraviolet light which inhibits other forms of growth.

Procedure

1. Prepare a spore suspension of *Penicillium*.
2. Using a counting chamber, determine the number of spores in 1 ml of this suspension.
3. Pour 2 Petri plates of nutrient agar.
4. Place 1 ml of the spore suspension on each of the agar surfaces.
5. Cover the whole surface of each plate by gently twirling the plates.
6. Set the plates on one side for about an hour for the spores to settle on the agar.
7. Irradiate one plate with short wave ultraviolet light for 40 seconds. To do this the lamp must be switched on some minutes before irradiating the spores and **the lid of the plate must only be removed when the plate is actually under the lamp.**

WARNING. Short wave ultraviolet light is dangerous to the eyes if looked at directly. Ideally an ultraviolet tube fixed on a stand and facing down on to the bench should be used.

8. Replace the cover to the plate before removing from under the lamp.
9. Incubate the plates for a week at 27°C.
10. At the end of this time, count the number of colonies which have grown on the treated plate and on the untreated plate.

11. Observe the appearance of colonies on the normal plate and on the treated plate. Look for abnormalities on the treated plate.

Question
What percentage of the colonies on the treated plate are abnormal?

Further Work
Try subculturing mutant colonies to check that it is an actual mutant and not merely an abnormality due to an unknown deficiency in the agar.

Repeat the original experiment using a longer period of irradiation. Does the time for irradiation affect the number of mutants formed? If time permits take sufficient readings to plot graphs of

(a) Irradiation time against the percentage surviving.
(b) Irradiation time against the mutants appearing.

Requirements
Apparatus and Reagents
Ultraviolet lamp producing light wavelengths shorter than 3100Å

2 Petri dishes, counting chamber, Nutrient agar

Biological
Penicillium spore suspension

BIOCHEMISTRY

Hydrogen ion concentration

The relative hydrogen ion concentration, or pH of the external environment is one of the most important factors concerned in the ecology of aquatic plants and animals. The distribution of many species may be altered by a change in the pH. Slight changes in the environment may have an adverse effect on the internal physiology of an animal.

The hydrogen ion concentration has a profound effect on the functioning of proteins, particularly enzymes. Each enzyme has an optimal pH at which it functions at its maximum rate, and there are mechanisms in animals which prevent too great a fluctuation of the hydrogen ion concentration.

Ionic product of water

Pure water is only very slightly ionised:
$$H_2O \rightleftharpoons [H^+] + [OH^-]$$
The amount of water ionised is about 1 gram molecule in 10 million litres.

The Law of Mass Action. The velocity of a chemical reaction is proportional to the product of the active masses of the reacting substances.

The degree of ionisation or dissociation of a weak electrolyte such as water can be determined by measurement of its electrical conductivity. According to the law of mass action,

$$\text{the equilibrium value, } K = \frac{[H^+][OH^-]}{[H_2O]}$$

or, since the concentration of undissociated water is constant,

$$\text{Ionic product of water, } K_w = [H^+][OH^-]$$

At 25°C. $$K_w = 1 \cdot 02 \times 10^{-14}$$

and therefore $[H^+] = [OH^-] = 10^{-14} = 10^{-7}$ gram-ion per litre.

Hydrogen ion index or pH

In all neutral solutions in water the hydrogen and hydroxide ion concentrations are equal $[H^+] = [OH^-]$

But since $[H^+][OH^-] = K_w = 10^{-14}$ is also true for **all solutions** (acid, alkaline or neutral), the hydroxide ion concentration is always given by

$$[OH^-] = \frac{K_w}{[H^+]} = \frac{10^{-14}}{[H^+]}$$

at 25°C, and alkaline solutions may be regarded as having a hydrogen ion concentration smaller than 10^{-7}, i.e. as having larger negative exponents. The whole range of reaction from acid to alkaline may thus be given in terms of the hydrogen ion concentrations or exponents.

Instead of the hydrogen ion concentration it is usual to give what is called the pH value, where the pH $= -\log H^+$, i.e. pH is the exponent of the hydrogen ion concentration (expressed as a power of 10) with the sign changed. In neutral solutions

$$pH = -\log 10^{-7} = 7.$$

In acid solutions the pH is less than 7, and in alkaline solutions it is greater than 7.

The biologist is less concerned with the theoretical aspects of pH, than with the practical implications. He is perhaps most interested in making measurements of the pH of the environment of any organism, or of its body fluids. In many experimental situations the biologist may wish to control the pH.

Methods of measuring pH

There are two very different methods of estimating the pH of a fluid in the laboratory or field, electrometric and colorimetric. The electrometric method, using a pH meter is the most accurate and reliable. Medical pH meters may measure to 0·002 of a pH unit, and the cheaper instruments which are finding their way into school laboratories are capable of measuring to 0·1 or 0·05 of a pH unit. This degree of accuracy is sufficient for any school needs. The colorimetric methods are less accurate, measurements can only be made within 0·3 of a unit, and this requires subjective judgement. A further disadvantage of the colorimetric methods is that biological fluids are themselves often coloured. Colorimetric methods are also affected by the presence of certain proteins in solution, giving rise to a 'protein error'.

pH meters

Fundamentally, these are instruments made up of a galvanometer, with its scale calibrated in pH units, and a Wheatstone bridge. The instrument measures an e.m.f. or potential difference which arises in the probe unit. The probe contains a glass bulb, filled with acid of known pH and an e.m.f. is set up because H^+ ions tend to migrate through the glass to the unknown solution. The greater the difference in the concentration of hydrogen ions on the two sides of the glass membrane, the greater the e.m.f. produced. The probe electrode is connected to one side of the Wheatstone bridge, the other side being connected to a standard reference electrode (a calomel electrode in the cheaper instruments). Very few hydrogen ions actually pass through the glass, but it is bad practice to leave the electrode in strong acids or alkalis for any length of time. Each time the meter is used it is necessary to standardise the Wheatstone bridge. This is done by placing the probe in a buffered solution of known pH, after the instrument has had a suitable period to warm up. If the pH is not accurately recorded on the scale the symmetry control is adjusted. (This control may be labelled symmetry, zero, or balance.) For accurate readings it is essential to standardise the instrument on a buffer in the range to be tested. For school work 3 buffers are sufficient, pH 4·0, 7·0 and 9·2. These may be purchased in tablet or sachet form to make 100 or 1,000 ml working solution. **Note. Distilled water cannot be used** in the place of pH 7·0 buffer because it may have a pH as high as 5·0 (due to the presence of carbonic acid from atmospheric carbon dioxide).

Setting up and using the pH meter

The instruments which are likely to be used in schools have two or three controls, the **symmetry control** (already described), a **2-position switch** which connects the circuit to the probe electrode or calomel electrode, and a **temperature compensator.** This latter control is only placed on the more expensive instruments. With the cheaper instruments most likely to be met in a school a correction chart is usually provided and the basic calibration is for 20°C. Provided solutions are allowed to assume ambient temperature before measuring pH correction will not be necessary for the experiments given in this book.

1. Switch on the instrument and allow 10 minutes to warm up.
2. Standardise with a buffer in the range required.
3. Rinse electrode in distilled water.
4. Immerse the probe in the test liquid.
5. After making the reading switch 2-way switch to 'Calomel', or '0'.

A moderately priced meter suitable for schools is produced by Analytical Measurements Ltd, The Quadrant, Richmond, Surrey. A new instrument for schools is being developed for the Nuffield 'A' Level Course and will be marketed by Phillip Harris Ltd.

Colorimetric methods

These depend upon the property of a group of weak, organic acids to change their degree of dissociation with the pH over a limited range, usually 2 pH units. A further property of these acids is that colour of the undissociated acid differs from the anion—with different degrees of dissociation, different proportions of acid and anion will be present giving a colour range. The colours produced on contact with a solution containing hydrogen or hydroxide ions can be compared against a standard chart. With this method it is essential to use a **wide range** or universal indicator in the first instance. If more accurate results are needed the appropriate **narrow range** indicator is selected. The two most useful methods of using indicators of this type are the **capillator method** and the **pH test paper method.**

Test papers

These consist of strips of paper impregnated with indicator. A colour chart is normally printed on the cover of the booklet.

1. Remove a strip from a wide range book.
2. Using a capillary tube (melting point tube), place a small spot of test liquid on the paper. The spot should be as small as possible otherwise the indicator is leached out and travels to the solvent front as the spot spreads, giving a dark ring with a lighter centre which cannot be compared satisfactorily with the standard colours.
3. Select a narrow range paper and repeat the test.

Accuracy is not likely to be better than 0·3 of a pH unit.

Capillator

A small amount of the test liquid is mixed with a small amount of indicator and then drawn up into a capillary tube. The capillary tube is then compared with a range of identical tubes containing standard colours until a match is found. As with papers it is necessary to use both wide and narrow range indicators unless the approximate pH is known with certainty. The capillator method is accurate to 0·2 of a pH unit.

Experiments

1. Measure the pH of the three solutions provided using
    (*a*) Test Papers
    (*b*) Capillators
    (*c*) pH Meter
2. Obtain a sample of the body fluid of an anaesthetised earthworm by passing a capillary tube obliquely through the body wall so that it enters the coelom. Use test papers.
3. Measure the pH of the contents of the intestine and the stomach of the mouse or rat provided. Use test papers.
4. Centrifuge a small sample of blood from the mammal provided and determine the pH of the plasma.
5. Test human saliva with capillator and test papers.

Requirements

Apparatus

Wide range papers	Narrow range capillators
Narrow range papers	Capillary tubes (melting point tubes)
Wide range capillators	pH Meter if available

Reagents

3 buffered solutions of different pH (See table in experiment on influence of pH on enzyme activity page or use buffer tablets)

Biological

Anaesthetised earthworm
Freshly killed mouse or rat

Buffer systems

It is essential that the biologist has an elementary understanding of buffer systems as many of the reactions which he investigates are buffered in nature. Furthermore the biologist, when designing an experiment, has to contend with many variable factors which affect a single reaction—one of these factors, the pH, or hydrogen ion concentration has a marked effect on the behaviour of proteins. It is essential, therefore, in many experiments, to stabilise the pH by the use of a buffered system.

A buffer system is a solution to which it is possible to add either hydrogen ions or hydroxide ions without affecting the pH of the solution. Buffer systems usually consist of a weak base and its salt with a strong acid, or a weak acid and its salt with a strong base. Let us consider a buffer system containing acetic acid and sodium acetate. In solution the components of this system will behave thus:

$$NaAc \rightleftharpoons Na^+ + Ac^-$$
$$HAc \rightleftharpoons H^+ + Ac^-$$

The acid being partially dissociated and the salt completely dissociated.

If additional H^+ ions are added to such a system, they will combine with the negatively charged anions to form undissociated acetic acid. If additional hydroxyl ions are added to the system, they will combine with H^+ ions and be removed as water. The hydrogen ions removed from the system are then replaced by further dissociation of the acid.

Such a system therefore has the ability to 'mop-up' both hydrogen ions and hydroxide ions, thus resisting any change in pH.

Biological implications

As slight shifts in the hydrogen ion concentration have marked effects on the functioning of proteins, including enzymes, it is not surprising that many biological systems employ buffers. A certain amount of buffering probably occurs in all cells and the most commonly occurring biological buffer systems are probably the bicarbonate/carbonic acid system, and the phosphate buffer system. Protein systems may also act as buffers and approximately one sixth of the buffering power of mammalian blood may be accounted for by plasma proteins.

Bicarbonate/carbonic acid buffer

Sodium is the chief cation present in mammalian plasma and may be regarded for the purposes of this argument as being present as dissociated sodium bicarbonate. As respiratory carbon dioxide dissolves in the plasma to form carbonic acid we have a buffer system

$$CO_2 + H_2O \rightleftharpoons H_2CO_3$$
$$NaHCO_3 \rightleftharpoons Na^+ + HCO_3^-$$

The control of blood plasma pH is tied up with the removal of respiratory carbon dioxide in the lungs

$$H_2CO_3 \underset{\text{carbonic anhydrase}}{\overset{\text{carbonic anhydrase}}{\rightleftharpoons}} CO_2 + H_2O$$

This is an excellent example of a homeostatic system (maintenance of the *status quo* of the internal environment). All homeostatic systems involve the detection of change, followed by a reaction to compensate the change, thus restoring the equilibrium. It is a feedback system, in this instance the anterior part of the medulla (respiratory centre) detects slight changes in the concentration of the carbonic acid in the plasma and an increase in the rate of removal of carbon dioxide from the lungs is initiated.

Phosphate buffer

Phosphoric acid forms two salts which are very frequently present in biological systems. Monosodium dihydrogen phosphate, (NaH_2PO_4) dissociates to give $H_2PO_4^-$ ions and can act in the same way as a weak acid by acting as a hydrogen ion donor to remove any additional hydroxide ions as water. The other salt, disodium hydrogen phosphate (Na_2HPO_4) completely dissociates to give HPO_4^- ions which possess a strong affinity for hydrogen ions which they remove from the system. The presence of these two salts gives a buffer system

$$NaH_2PO_4 \rightleftharpoons Na^+ + H_2PO_4^-$$
$$Na_2HPO_4 \rightleftharpoons Na^+ + HPO_4^-$$

Protein buffer

Proteins possess both basic and acidic groups ($^-NH_2$ and ^-COOH). Such a molecule is able to ionise in two different ways, depending on the pH of its surroundings. In acid conditions the $^-NH_2$ radical acts as a hydrogen acceptor, becoming $^-NH_3$. In alkaline conditions the ^-COOH group acts as a hydrogen donor, becoming COO^-.

The buffering power of proteins varies considerably. In most protein molecules the majority of $^-NH_2$ and ^-COOH groups are linked in peptide bonds $^-CO.NH^-$ and are unable to play a role in buffering. The only free $^-NH_2$ and ^-COOH groups occur at the end of the peptide chains. However, some proteins possess the amino acid, histidine, which contains additional $^-NH_2$ and ^-COOH groups in its molecule. As these groups are not required in peptide

100

bonds they are free to act in buffering systems. Haemoglobin contains 36 histidine units in its molecule and is therefore an important buffer.

Buffering effects of the environment

Dissolved substances in the environment may act as buffer systems and may have a marked influence on the physiology of organisms in the environment. Small organisms such as bacteria and yeasts are particularly influenced by environmental changes.

Sea water is an example of an external environment with considerable buffering powers.

Preparation of phosphate buffer

1. Make up a M/5 solution of anhydrous disodium hydrogen phosphate (Na_2HPO_4). (28·4 g/litre)
2. Make up a M/5 solution of potassium dihydrogen phosphate, (KH_2PO_4). (27·22 g/litre)
3. These stock solutions keep well if stored in a refrigerator. If a refrigerator is not available they should be made up just prior to use.
4. When a phosphate buffer of any pH is required, it may be prepared by mixing the stock solutions as shown in the table below. These are proportions (parts per 100), and if buffers of pH intermediate to the values given are needed, the proportions can be easily obtained from a graph with parts of Na_2HPO_4 on one axis and pH on the other. The parts per 100 of Na_2HPO_4 can be read direct from the graph and the value for KH_2PO_4 is then obtained by subtracting this figure from 100.

pH	M/5. Na_2HPO_4	M/5. KH_2PO_4
5·3	2·5	97·5
5·6	5	95
5·9	10	90
6·2	20	80
6·5	30	70
6·65	40	60
6·8	50	50
7·0	60	40
7·15	70	30
7·4	80	20
7·75	90	10
8·05	95	5

This buffer is very useful for school work and can be prepared very cheaply. Commercial buffer tablets are also available and save a considerable amount of time in the preparation of solutions, they are, however, more expensive and a wide range has to be stocked.

An experiment to demonstrate the effect of buffer action

Procedure

Titration A

1. Place 20 ml N/10 acetic acid in a 100 ml beaker.
2. Add a few drops of phenolphthalein as an indicator and titrate against N/10 sodium hydroxide, adding 1 ml at a time and shaking.
3. After the addition of each ml of sodium hydroxide, the glass electrode of the pH meter should be inserted in the beaker and the reading recorded. **Do not use the glass electrode (probe unit) as a stirring rod. It is extremely delicate.**

Titration B

1. Prepare a buffer system of acetic acid/sodium acetate by dissolving 4·1 grams of sodium acetate in 1 litre of N/10 acetic acid.
2. Place 20 ml of the buffer solution in a 100 ml beaker and add a few drops of phenolphthalein.
3. Record the pH using the pH meter.
4. Add N/10 sodium hydroxide, recording the pH after each 1 ml addition.

Results

1. Plot a graph with pH on the vertical axis and mls. NaOH on the horizontal axis. If the titration curves of both titrations are plotted on the same graph, the buffer effect should show clearly.
2. Subtract successive readings from each other to obtain the **change in pH** for each 1 ml of sodium hydroxide added. These results may be plotted graphically or expressed as a comparative table for the two titrations.

N.B. If a pH meter is not available this experiment may be carried out by an alternative method which is rather more laborious. Proceed as above but omit the phenolphthalein and record the pH after each addition of sodium hydroxide by using pH test papers. This method is of course much less accurate than using a meter as the sensitivity of the papers is only about one tenth of the sensitivity of a meter.

Requirements

Apparatus

50 ml burette	pH test papers
Burette stand	White tile
100 ml beaker	Burette funnel
pH meter or wide	Balance
and narrow range	20 ml pipette

Reagents

Phenolphthalein (optional)
N/10 acetic acid
N/10 sodium hydroxide
Sodium acetate
N.B. As the grease used on the tap of a burette is affected by NaOH, burette should be cleaned and regreased after use

Chemical tests for proteins

There are a number of chemical tests which can be used to confirm the presence of protein in biological material. These tests depend upon the reaction between a radical or group in the protein molecule and the test reagents, to form a coloured compound. As proteins may vary in the proportions of the reactive group which they possess, the intensity of the colour formed in a test will vary with the protein tested.

A positive reaction in one of the tests listed below cannot be taken as proof that a protein is present as the individual tests only confirm the presence of a particular group which may occur in materials other than proteins. Each of the tests given below confirms the presence of a different group and a material which gives a positive reaction in all of the tests may be regarded with confidence as being a protein.

Millon's test

Millon's reagent is a solution of mercuric nitrate in nitric acid. It will react with any phenolic compound in which the 3 and 5 positions are unsubstituted. Proteins give a red coloration with Millon's reagent because an unsubstituted phenolic ring occurs in the amino acid tyrosine.

Phenol Ring
(3 and 5 positions arrowed)

Tyrosine part of protein molecule

Millon's test can be carried out on solutions and solids.

Biuret test

Coloured complexes may be formed by the reaction between solutions of copper salts and adjacent pairs of $-CONH_2$ (Carbamyl groups). Similar reactions occur with $-CSNH_2$; $-C(NH) NH_2$ and $-CH_2NH_2$ groups. The pairs of carbamyl groups may be joined directly to each other, or through a single atom of carbon or nitrogen.

Oxamide

Biuret

103

A portion of a protein molecule contains the following configuration which will give a positive biuret test.

It is probable that the coloured complex formed has a ring structure.

Xanthroproteic test

Aromatic rings may be nitrated with concentrated nitric acid to give nitro-compounds which give an orange or yellow coloration when in alkaline solution. The amino acids tryptophan and tyrosine possess aromatic rings and cause proteins to give a positive reaction to the test.

Ninhydrin test

Ninhydrin (1, 2, 3 — indanetrione hydrate) reacts with amines, particularly primary amines to give coloured products. —NH or —NH$_2$ groups are involved in the reaction. The ninhydrin test gives positive results with amino acids, peptides, peptides and proteins. In these cases adjacent carboxyl and amino groups appear to be reacting. Simple amines without carboxyl groups, such as ammonium compounds will also give a positive reaction.

The coloration given by different amino acids in the ninhydrin test enables an experienced worker to judge which amino acids are present in a chromatogram which has been developed with ninhydrin.

Ninhydrin is a carcinogen. Skin contact and the inhalation of aerosol particles must be avoided. Aerosols must be used in a fume cupboard.

Procedure

Millon's test

1. Dilute 1 ml Millon's reagent with 2 ml distilled water.
2. Add small quantity of the test material either as a solid or as a solution.
3. Heat gently. If the test is positive solids will become tinged with red or the solution turns red.

Test albumen, milk, cheese and phenol. (Take great care when

104

handling phenol as it is very corrosive and must be kept off the skin.)

Biuret test

1. Take 2 ml of the test solution.
2. Add 2 ml 10% sodium hydroxide solution and mix well.
3. Add 0·5% copper sulphate solution drop by drop, shaking well between each addition. A coloration from pink to blue, but usually purple or violet is positive. (The positive coloration in this test must not be confused with the blue precipitate of copper hydroxide which is formed if excess copper sulphate solution is added.)

Test albumen, milk, cheese and biuret.

Xanthroproteic test (use extreme care)

1. Take 2 ml. of the test solution.
2. Add 1 ml concentrated nitric acid which will throw down any protein present as a white precipitate.
3. Heat carefully. The white precipitate turns yellow as nitration occurs.
4. **Leave the mixture to cool** and then to a **few drops** of it, add excess 10% sodium hydroxide solution. The yellow coloration changes to orange.

Test albumen, milk, cheese and a very dilute solution of phenol.

Ninhydrin test

1. Take 5 ml of dilute test solution.
2. Check that the pH of the solution lies between 5·0 and 7·0.
3. Add 0·5 ml 0·1% ninhydrin solution.
4. Boil for 2 minutes.
5. Cool. The coloration, usually blue or purple, appears as the solution cools.

Test albumen, milk, cheese and ammonium acetate.

Requirements

Apparatus

Test tubes
Boiling tubes

Test tube racks
Test tube brushes

Reagents

Millon's reagent
10% sodium hydroxide solution
0·5% copper sulphate solution
Concentrated nitric acid
0·1% ninhydrin solution
Albumen
Milk
Cheese
Phenol
Ammonium acetate

The solubility of proteins

The chief factors involved

1. The spatial arrangement of the peptide chains and the internal nature of the molecules. If this is disorganised, the protein becomes denatured. This leads to the unfolding of the tightly coiled protein chain and hence to the disorganisation of the internal structure. Now the protein is no longer globular, the solubility is decreased. The precipitate is soluble in acids and alkalis. Also some previously concealed groups are now exposed at the surface of the molecule and therefore they can be acted on by chemical agents, e.g. the thiol groups (—SH). Note that denaturation and coagulation are not the same process. Denaturation precipitates the protein fibrils; if these are then heated they will coagulate. After coagulation the precipitate is no longer soluble in acids and alkalis. A great many proteins do not coagulate on heating, but they are precipitated when denatured, e.g. casein and gelatin.

2. The colloidal nature of protein solutions.
   Hydrophilic colloids have two stability factors which keep them in solution by preventing the formation of precipitates by keeping the individual particles apart. These factors are the water molecules which surround the protein and the electrical charges which some proteins carry.

3. The amphoteric nature of the proteins.
   Proteins can form salts of two types. A protein forms **zwitterions**
   thus $NH_2 . X . CO . OH = {}^+NH_3 . XCO . O^-$
   At the isoelectric point (see the next experiment) the protein is in equilibrium with its neutral zwitterions.
   Protein anions can combine with cations and protein cations can combine with anions.

Experiments

1. Denaturation

(a) Place 3 ml of egg white in each of two test tubes, A and B. Boil B for 10 seconds. Cool.
   Add 3 drops of bromocresol green to each tube.
   Add 0·1 N HCl drop by drop to each tube until the blue colour becomes green. Mix after each drop. The pH is now below the isoelectric point of albumen and globulin.
   In B a precipitate of denatured albumen and globulin forms as the acid is first added but this later dissolves as more is added.
   In A no precipitate forms. Because the protein has not been

106

denatured, whereas in B boiling denatured the protein and the acid is redissolving the precipitate.

(b) Add 0·5 ml of 0·1 N HCl to 5 ml of egg white.

Mix and heat to boiling.

The denatured protein is soluble in dilute acid and so the solution remains clear.

Add 3 drops of chlorophenol red.

Add 0·1 N NaOH drop by drop until the colour is faintly red. Mix thoroughly after each drop.

Add 1 drop of 1% acetic acid to bring the solution near the isoelectric point (pH 5·4).

The denatured protein is now completely precipitated.

After shaking the precipitate to thoroughly disperse it in the solution, pipette 1 ml of it into each of 4 test tubes, A, B, C and D.

Add 2 ml of water to A and shake.

Add 3 drops of 0·1 N NaOH to B.

Add 3 drops of 0·1 N HCl to C.

Boil D.

In A the precipitate does not dissolve; in B and C the precipitate dissolves; in D a coagulum is formed, which will not dissolve on the addition of 3 drops of 0·1 N HCl.

2. Increased accessibility of thiol groups

Set up a series of test tubes as follows:

Tube 1. 0·2 ml 2,6-dichlorophenol indophenol; 4·8 ml de-ionised water.

Tube 2. 1·0 ml casein; 0·2 ml 2,6-dichlorophenol indophenol; and 3·8 ml of de-ionised water.

Tube 3. As tube 2.

Keep tubes 1 and 2 at room temperature; place tube 3 in a boiling water bath for 5 minutes.

Cool in a beaker of cold water.

2,6-D is a blue dye which can be reduced to form a colourless solution,

R—SH + HS—R + dye coloured
R—S—S—R + leuco dye colourless.

Compare the intensities of dye in the three tubes.

3. Precipitation by dehydrating agents

Place 2 ml of 96% ethyl alcohol into each of 2 test tubes, A and B.

Add 5 drops of blood plasma to each tube.

A dense precipitate will form.

Add 10 ml of water to A and mix. The precipitate will dissolve. Leave B for 30 minutes and then add the water. It will not dissolve.

107

4. Precipitation with concentrated salt solutions

(a) Dilute blood plasma 10 times.

Add 1 ml of this diluted plasma to an equal volume of saturated ammonium sulphate. A precipitate of globulin forms.

Add 2 ml of water, mix. The precipitate dissolves.

(b) Add 10 ml of saturated ammonium sulphate to 10 ml of the diluted plasma.

Mix thoroughly and then filter off the precipitate.

Continue filtering until a clear filtrate is obtained.

Add 1 drop of 1% acetic acid to 2 ml of the filtrate.

A coagulum appears due to the albumen (globulin has been filtered off).

(c) Saturate 2 ml of the filtrate by adding a small spoonful of ammonium sulphate crystals and shake.

Continue adding crystals until some remain undissolved.

A precipitate of albumen will form which can be redissolved in water.

Note that globulin has been precipitated with half saturated ammonium sulphate and albumen requires full saturation to precipitate it.

5. Precipitation with specific anions and cations

(a) Place 1 ml of egg white in each of 3 test tubes.

Add 3 drops of 2% copper sulphate to one tube.

Add 3 drops of 2% lead acetate to another.

Add 3 drops of 4% mercuric chloride to another.

(b) Add 1 ml of egg white to 5 drops of 10% trichloroacetic acid.

Requirements

Apparatus

Test tubes
Boiling water bath
Filter funnel and paper

Biological

Egg white
Blood plasma

Reagents

Bromocresol green
0·1 N HCl
Chlorophenol red
0·1 N NaOH
1% acetic acid
2,6 dichlorophenolindophenol
Casein
96% ethyl alcohol
Saturated ammonium sulphate solution
Ammonium sulphate crystals
2% copper sulphate
2% lead acetate
4% mercuric chloride
10% trichloroacetic acid

Determination of the isoelectric point of a protein

Proteins are at their least soluble when there is no net overall charge on their molecules, that is when there are equal numbers of positive and negative charges. The ionisation of a protein molecule depends upon the surrounding pH and at one specific pH the protein is at its lowest point of solubility. This is the point at which the protein in question is electrically neutral and is known as the **isoelectric point.**

At their isoelectric point, some proteins, e.g. casein and edestin, are almost insoluble, whilst others such as gelatine still remain soluble but may be easily precipitated by the addition of ethanol.

Procedure

1. Make up a solution of casein in sodium acetate as follows:

    (a) Place 0·25 g of pure casein in a 50-ml volumetric flask.

    (b) Add approximately 20 ml distilled water and **exactly** 5 ml N sodium hydroxide solution. Shake until the protein is dissolved.

    (c) Add **exactly** 5 ml N acetic acid to neutralise the alkali.

    (d) Dilute to 50 ml with distilled water to give a solution of casein in 0·1 N sodium acetate.

2. Prepare a series of tubes as shown in the table below.

Tube number	1	2	3	4	5	6	7	8	9
ml distilled water	8·38	7·75	8·75	8.5	8	7	5	1	7·4
ml 0·01 N acetic acid	0·62	1·25	—	—	—	—	—	—	—
ml 0·1 N acetic acid	—	—	0.25	0·5	1	2	4	8	—
ml N acetic acid	—	—	—	—	—	—	—	—	1·6

3. Add 1 ml of the casein in 0·1 N sodium acetate to each tube. The addition should be made quickly, blowing the casein solution out of the pipette if necessary. Shake each tube and record any turbidity or precipitate present immediately after the addition.

4. After 10 minutes, and again after 30 minutes, record the degree of turbidity or precipitation present in each tube. The results should be recorded in a table such as that given overleaf. Indicate a precipitate in the table with an x and show degrees of turbidity by the use of +, ++ or +++. Indicate the complete absence of turbidity in the table with 0.

Note. Each tube is a buffered solution of acetic acid/sodium acetate and its pH has been carefully calculated. Accuracy in making up the tubes and solutions used in this experiment is essential if a reliable result is to be obtained.

Tube number	1	2	3	4	5	6	7	8	9
pH	5·9	5·6	5·3	5·0	4·7	4·4	4·1	3·8	3·5
Turbidity Immediate After 10 mins After 30 mins									

If a pH meter is available the accuracy of the buffered solutions can be checked.

Results

From the table it is possible to estimate the isoelectric point of the casein within 0·3 of a pH unit. If the school possesses a colorimeter (Absorptiometer), very precise results can be obtained and the turbidity can be plotted against pH on a graph and the precise isoelectric point read off from the line obtained.

Requirements

Apparatus

Test tubes
Test tube rack
10 ml graduated pipette
1·0 ml graduated pipette
50 ml volumetric flask

Optional apparatus

pH meter
Colorimeter and tubes

Reagents

Pure casein
Distilled water
N Sodium hydroxide solution
N Acetic acid solution
0·1 N Acetic acid solution
0·01 N Acetic acid solution

Chromatography

This is a technique for the separation of a mixture of solutes in which separation is brought about by using the different rates of movement of the individual solutes through a porous medium under the influence of a moving solvent.

This depends on multiple partition or adsorption–desorption process. During the passage through the chromatographic system, small differences in the partitioning behaviour of each component of the mixture are multiplied many times.

Partition chromatography is carried out on sheets of filter paper or on very thin layers of kieselguhr, moist silica gel or powdered cellulose. The porous medium acts as a support for water, which acts as the partitioning agent. Filter paper consists of many cellulose fibres which trap a certain amount of water. Consider that each fibre together with its trapped water constitutes a 'cell'. Separation of the solutes is brought about by the partition of these substances between the water in the 'cells' and the solvent which flows over the 'cells'. The water in the 'cells' remains stationary. The distance moved by a solute will depend on its relative solubility in the stationary water and the solvent. For instance a substance which is only soluble in the solvent will move as far as the solvent moves, whereas a substance which is insoluble in the solvent will remain on the point of the porous medium to which it was applied.

Other factors such as adsorption will affect the movement. The partition of a substance between two immiscible solvents is not affected by other substances.

The mixture of the solutes is applied at the origin and the paper is placed with one end in the solvent such that the origin is clear of the solvent. The solvent rises up the paper by capillarity. When it reaches the 'cell' on which the solute mixture has been spotted, the separation of the solutes begins. The solvent then containing an amount of the solute mixture moves on to the next 'cell'. Fresh solvent will come into contact with the first 'cell' and a further partition of the remaining solute mixture will take place. This is the countercurrent distribution and it continues as the solvent moves along the paper. The process is repeated a great many times and finally a sharp separation of the solutes is achieved.

Paper chromatography has made possible the speedy chemical analysis of complex organic chemicals. It has also made possible the separation and identification of microgram quantities of substances in unknown mixtures. ($1\mu g = 0.000001$ g $= 10^{-6}$ g.)

Paper chromatography

Procedure

Solvent
Prepare the required solvent in a measuring cylinder, and mix thoroughly by inverting several times. Use the smallest cylinder possible.

Application of sample to paper

1. Use 25 cm × 25 cm Whatman number 1 filter paper.
2. Draw a line with a pencil across the paper, 2·5 cm from one edge.
3. Beginning 2·5 cm from the end of this line, place pencil crosses on the line every 2·5 cm. Each will serve as the origin for a solution of the test substance, or for a known substance against which it is being compared.
4. With the pencil indicate the nature of the substance to be placed on each origin under the crosses.
5. Place the paper on the bench with the origin line overhanging the edge.
6. Spot a drop of each substance on to the paper with a piece of fine capillary tubing. A clean piece of tubing must be used for each origin.
   Alternatively a 4 mm diameter platinum wire loop can be used to apply the sample. The loop must be washed in water to remove metal salts and flamed to remove organic substances between each application.
7. The spots should be about 5 mm diameter and definitely not greater than 8 mm diameter.
   Drawing the capillary tube out into a fine point helps in getting small spots.
8. If a greater concentration is required on each origin than is given by one application, the first application must be allowed to dry before adding the second.
   A hair drier can be used to accelerate drying.

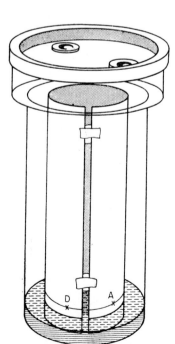

Operation

1. Bring the two edges of the paper together to form a cylinder, and fasten with two plastic, tongued clips. These are made to keep the two edges apart as it is essential that they do not touch. The line containing the origins must be in the horizontal plane.
2. Use an air-tight tank for the run. This can either be a Shandon Unikit tank or belljar, crystallising dish and ground-glass plate.
3. Place sufficient solvent in the tank or the crystallising dish to cover the bottom centimetre of the filter paper when this is placed in the solvent.
4. Place the lid on the tank or the belljar over the crystallising dish and leave for a short time for the volatile solvents to saturate the atmosphere.
   Before using a container wash it out with a small amount of the solvent.
5. Lower the paper cylinder, spotted end first into the solvent. **Take care that the lower edge of the paper is horizontal as it touches the solvent** and is then lowered further into it. If you do not take this precaution, the solvent front will slope and subsequently the measurement of R_f values will be difficult. If it is

necessary to move the tank while the separation is in progress, it is also important that the solvent front is kept horizontal.

Make sure the paper is not touching the glass sides of the container.

6. Place the lid on the tank or the belljar over the crystallising dish. If using the Shandon Unikit tank, check that the two holes in the lid are closed with the two polythene stoppers supplied with the kit.

   If using the belljar arrangement, check that the jar is in contact with the ground glass side of the plate and not the smooth side.

7. If the solutes are coloured, the separation can be followed visually and the separation can be ended when an adequate separation has been obtained.

   If they are not coloured, you must become familiar with the solutes being studied, and it may be necessary to run additional chromatograms for different times to achieve the desired degree of separation.

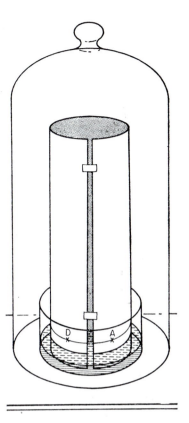

Drying the chromatogram

1. When the chromatogram has run for the desired time, remove it from the tank and open it out.
2. Mark the solvent front with a pencil.
3. Dry it as rapidly as possible, either by holding it over a low bunsen flame or over a photographic film drier. If a hair drier is available this is the best method.
   WARNING. Care must be taken with inflammable solvents.
   If a fume cupboard is available this ought to be used.
   The process of drying the chromatogram, fixes the solutes at the sites they have reached.

Locating the separated substances

1. When the substances are coloured no further treatment is needed. If they are colourless, the dried paper must be treated with a locating reagent. This is a substance which reacts with them to produce a product which is visible.
2. To do this fill a dipping tray with the locating reagent. The

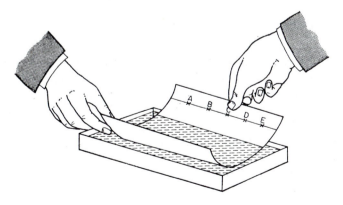

dipping tray is a shallow tray of such dimensions that a chromatogram can be drawn through the reagent it contains without touching the sides.

The paper should be dipped once only, from one end to the other, and then immediately removed from the tray.

Alternatively the paper can be sprayed with the aid of an atomiser or a special chromatographic spray bottle. A scent spray bottle is suitable. **Ninhydrin spray must be used in a fume cupboard.**

3. A number of unsaturated organic compounds fluoresce. These compounds absorb ultraviolet light or light of short wavelength and emit light of longer visible wavelength. These compounds can therefore be seen under an ultraviolet light. Each compound emits light of a characteristic wavelength and hence of a characteristic colour. The colours can be used for identifying the compounds.

4. 'Labelled' radioactive compounds can also be used. They can then be detected on the chromatograms with the aid of a Geiger counter. This method has the advantage over chemical methods in that the substances on the chromatogram are not converted into other compounds, and so they can be removed from the paper for further study.

Two-way chromatography

In some cases one-way chromatography as so far described, only brings about a partial separation of the substances. In this case a second solvent at right angles to the first should be used.

When doing this start the first separation with only one spot instead of the series of spots at 2·5 cm intervals which are applied to the paper in one-way chromatography.

The one spot must be applied to a point 2·5 cm from one side and 2·5 cm from an adjacent side.

After using the first solvent, remove the paper from the tank and dry it. Then turn it through 90° and place it in the tank in the second solvent. Complete the procedure as for one-way chromatography.

Between each run the tank must be cleaned and dried.

Descending chromatography

So far ascending chromatography has been described. Alternatively the solvent can be allowed to descend down the paper. In this case the solvent is placed in a trough suspended across the top of the tank. Some solvent is also poured into the bottom of the tank to saturate the atmosphere. The spotted paper is suspended from the solvent in the trough. Anchor it in the trough with a glass rod. The spots on the paper must be at the trough end of the paper at the beginning of the run.

Serrate the edge of the bottom of the filter paper to allow the solvent to run off it more easily.

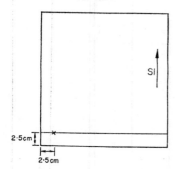
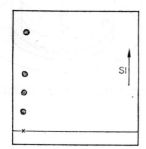
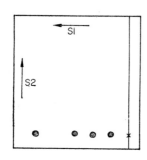
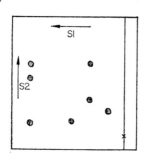

114

All the experiments in this book use the ascending technique, as this is the easier to set up. A disadvantage of this technique is that compounds which are relatively insoluble in the solvent, are often not separated completely. In this case the descending technique must be used so that the developing solvent can be allowed to run off the end of the paper under the influence of gravity. This increases the effective run and hence improves the separation.

Results

The movement of the substance relative to the solvent front is constant and characteristic of that substance. This is the R_f value.

$$R_f = \frac{\text{Distance moved by the substance}}{\text{Distance moved by the solvent front}}$$

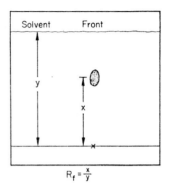

If the solvent front has run off the end of the paper, compare the movement of the substance with that of a reference substance, which must be chemically similar to the one being studied. This is the R_x value which may be defined as

$$\frac{\text{Distance moved by the substance}}{\text{Distance moved by the standard substance}}$$

Mannose is relatively insoluble and hence the solvent front may be allowed to run off the end of the paper. When investigating sugars, glucose is usually used as the standard substance.

These constants are usually expressed as percentages.

These values are as characteristic of a substance as its melting point, provided that the conditions are always reproduced.

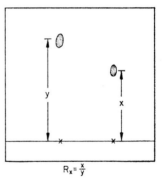

Requirements

Apparatus

Shandon Unikit Tank, lid with poly-thene stoppers
Alternatively a 7 cm diameter crystallising dish can be used to hold the solvent. This should be rested on a ground glass plate
Cover these with a belljar, 30 cm high to the shoulder
25 × 25 cm Whatman No. 1 filter paper
Plastic tongued clips. Shandon include suitable ones in their kit
4 mm diameter platinum wire loop or pieces of capillary tube
Ruler

Hair drier if one available, or photographic drier, or Bunsen
Fume cupboard if available
If descending chromatography is to be practised, you will also need a trough with anchoring glass rod
30 × 5 cm Whatman No. 1 filter paper
Shandon Unikit tank or some other suitable tank with supports for the trough must be used

Reagents

A suitable solvent
A suitable locating agent if one is necessary, or an ultraviolet lamp if one is necessary, or a radioactivity counter

115

The separation and identification of amino acids by paper chromatography

Note. In all work with amino acids, wash your hands thoroughly before handling the paper because perspiration contains amino acids. The paper should only be handled by the edges because finger marks show up as confusing coloured spots after treatment of the paper with locating agent.

Procedure

Materials to be investigated

This experiment is best done first with pure samples of amino acids and then with naturally occurring amino acids, such as those in fruit juices.

(a) Pure samples—Using a 25×25 cm piece of filter paper, place one drop of each solution of known amino acid and one drop of a mixture of all the amino acids on the paper.

(b) Fruit juices—Squeeze the juice from a fresh fruit such as a tomato and centrifuge or filter it.

Set 1 ml or more of the filtrate aside for chromatography. Treat a second 1 ml portion with 3 ml of 80% ethyl alcohol. This will precipitate the protein and hence make it possible to examine the fruit juice for free amino acids.

Centrifuge or filter the treated juice and use the filtrate for chromatography.

Mark seven origins on the paper; on origin 1 place 1 drop of untreated fruit juice; on origin 2 place 6 drops of treated juice; use each of the other origins for known amino acids, such as aspartic acid, asparagine, proline, leucine and lysine.

Solvent

Ethyl alcohol   80 parts by volume
Water   10 parts by volume
0·880 ammonia 10 parts by volume

Running time

At least $1\frac{1}{2}$ hours

Locating reagent

200 mg of ninhydrin dissolved in 100 ml of acetone
Store in a refrigerator

Results

Determine the R_f values of each substance after outlining its position with a pencil.

116

Questions

1. Is the R_f value of an amino acid affected by being run on the chromatogram as a single substance or as part of a mixture?
2. Are there any differences between the chromatogram of the treated juices and untreated juices? If there are any differences, can you offer an explanation for them?

Further Work

1. Investigate the changing amino acid content of fruits throughout the year and as a fruit matures.
2. Investigate fruits from a single family for similarities in the amino acid content.
3. If a 2-way chromatogram is prepared for the fruit juices, do any of the substances separate out further?
   For this purpose use the following solvent:

n-butanol	60 parts by volume
glacial acetic acid	10 parts by volume
de-ionised water	20 parts by volume

Requirements

(In addition to those needed for all chromatography experiments)

Apparatus

Filter funnel and filter paper or a centrifuge

Reagents

A collection of concentrated solutions of several amino acids, including the following: Aspartic acid, proline, asparagine, leucine, lysine
80% ethyl alcohol
Absolute ethyl alcohol

0·880 ammonia
Acetone
Ninhydrin
n-butanol
Glacial acetic acid

Biological

Fruits

An analysis of the amino acids present in a protein

The peptide linkages between the amino acids which are combined to form a protein may be broken by hydrolysis. The hydrolysis of the peptide bonds may be carried out with acids or alkalis but in practice it is found that the acid hydrolysis causes less destruction of the amino acids present. Tryptophan and cystine are destroyed by acid hydrolysis and in accurate work they are estimated by a separate means. In the fairly simple technique outlined in this experiment, some amino acid is inevitably lost but at least eleven of the constituent amino acids of casein can be demonstrated.

After hydrolysis of the protein, the constituent amino acids are separated by two-way paper chromatography.

Procedure

1. Place 10 mg casein in a 50 or 100 ml round bottomed flask.
2. Add 20 ml of 6 N hydrochloric acid.
3. Attach a Liebig condenser to the flask.
4. Reflux for 18 hours. As the apparatus must be left overnight, it should be stood on an asbestos sheet and an isomantle should be used as a heating source.
5. Evaporate excess acid in a waterbath, preferably in a fume cupboard.
6. Desiccate for 24 hours over soda lime. (Use a vacuum desiccator if one is available.)
7. Add 20 ml distilled water and warm gently to dissolve the residue.
8. Filter and evaporate the filtrate to dryness.
9. Dissolve the residue in 1 ml 10% aqueous isopropanol.
10. Spot the mixture on to a 25 cm square of chromatography paper and allow to dry thoroughly.
11. Place the paper in a chromatography tank, using 50 ml of a solvent system made up of 12 parts butanol, 3 parts acetic acid and 5 parts water.
12. Run overnight (15–20 hours).
13. Remove paper, dry thoroughly.
14. Turn paper through 90° and place in a second solvent system consisting of 50 parts phenol (**Handle with EXTREME CARE**), 12 parts distilled water, add 0·25 ml 0·880 ammonia per 50 ml just before use.
15. Run for 4 hours. Remove and dry paper.
16. Dip in ninhydrin solution, dry in air, then warm gently preferably on a hotplate or **high above a bunsen.**
17. Make a pencil ring around all spots on your chromatogram. Some spots fade when the paper is stored.

Requirements

Apparatus

Chromatography tank (Shandon Unikit is ideal)	50 ml measuring cylinder
	1 ml graduated pipette
50 or 100 ml round-bottomed flask preferably with ground-glass joint	Asbestos sheet
	Filter paper and funnel (or centrifuge)
Liebig condenser	
Isomantle and regulator	Water bath
Retort stand	Desiccator
Clamp and boss	Hotplate or bunsen

Reagents

Casein	Acetic acid (Glacial)
6 N hydrochloric acid	Phenol
Soda lime	0·880 ammonia
Isopropanol	Ninhydrin solution (200 mg in 100 ml
Butanol	acetone)

Paper electrophoresis

Proteins and amino acids migrate in an electric field except at the isoelectric point. At this point the net positive charge equals the net negative charge. The net charges vary with pH. Therefore two proteins or amino acids with different isoelectric points will migrate at different rates. They can therefore be separated by paper electrophoresis. The technique is most effective when the solution is buffered.

Procedure

Apparatus

Use the equipment shown in the diagram. Suitable equipment can be obtained from Shandon Scientific Company.
Alternatively a piece of filter paper can merely be suspended between two crystallising dishes, each containing buffer.
An electrode is placed in each of these dishes.
If using the Shandon equipment, use a piece of filter paper 30 × 10 cm
Shandon also supply a special power pack for this experiment.

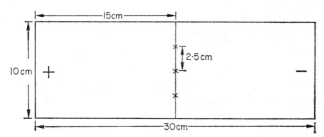

Operation

1. Draw a feint line across the filter paper, 15 cm from one end.
2. Indicate the individual origins with pencil crosses.
3. Mark the positive end with a + and the negative end with a —.
4. Clamp the paper support about 15 cm above the buffer compartments.
5. Wet the paper by dipping it in the buffer solution. Drain off the excess buffer by blotting it with clean filter paper.

119

6. Suspend the paper from the support so that each end is in a different buffer compartment. Each compartment is supplied with an electrode. Make sure that the paper is placed according to the + and − labelling.

7. Add buffer to each compartment until it is above the level of the electrode. In the Shandon apparatus there are four holes near the bottom of each pillar and the buffer must cover these holes as these make contact with the electrodes. In the Shandon apparatus about 100 ml of buffer is required.

There must be equal volumes in each compartment to prevent siphoning from one to the other via the paper strip.

8. If applying only a single spot of the substance being investigated to the paper, this should now be applied with a piece of capillary tubing.

Note. If applying more than one drop to an origin, this procedure must be modified thus:

Apply the spots to the dry paper with drying between each application.

Now wet the paper at both ends, leaving a dry strip in the middle. To do this, hold it at both ends and insert the paper in the buffer at a point about 2 cm from the line of the origins and draw the paper through towards the end of the paper.

Repeat this on the other end of the paper.

Blot the paper to remove the excess buffer, taking care not to wet the centre of the paper.

Place the paper on the support with the ends in the buffer.

With a piece of capillary tubing, wet the paper in the central dry region but do not wet the origins directly, merely allow the buffer to diffuse into the origins.

9. Insert the apparatus into the tank and cover with the lid.
10. Plug the electric leads into the electrodes.
11. Switch on the current.

Do not touch the apparatus until the current is switched off.
12. At the end of the experiment, remove the paper from the tank.
13. Blot the ends which were in the buffer.
14. Dry the paper as quickly as possible.
15. Locate the separated substance as described in the paper chromatography account.

Requirements

Apparatus

Chromatography tank
Electrodes. It is best to use those made by the Shandon Scientific Company
Power pack. This can also be obtained from the Shandon Scientific Company

A pair of electrical leads
The Shandon equipment includes the supports for the paper and the buffer compartments
30 × 10 cm filter paper

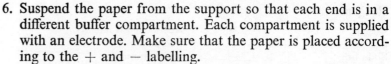

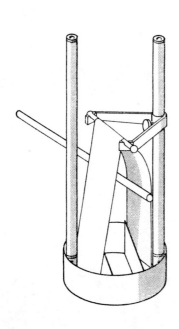

The separation of amino acids by electrophoresis

Procedure

This experiment is best done first with pure samples of known amino acids and then with naturally occurring amino acids.

(a) Pure samples: Use only one drop of each amino acid.
On one origin place one drop of a mixture of the acids.
(b) Fruit juices: Squeeze the juice from a fresh fruit.
Centrifuge or filter it to remove the solid material.
Then place one drop of each juice on separate origins and a drop of the amino acid mixture on another origin.

Buffer

This must be pH 6·1.
To prepare this solution add 0·8 ml of glacial acetic acid to 10 ml of pyridine and make the solution up to 250 ml with de-ionised water.

WARNING. There appear to be conflicting opinions about the desirability of having pyridine in the school laboratory. 1 p.p.m. in the atmosphere can cause headaches and nausea. If the liquid comes in direct contact with skin or eyes it can cause dermatitis and conjunctivitis respectively. It must therefore be handled with care and should only be used in a fume cupboard or by an extractor fan.

It has also been suggested that pyridine can cause sterility and that it is carcinogen. Opinions on the validity of this statement seem to differ widely and so in the absence of any firm evidence it is felt that this limited use of the chemical in the school laboratory, supported by sensible precautions, is justified.

Running time

1–1½ hours

Locating reagent

200 mg of ninhydrin dissolved in 100 ml of acetone
This must be stored in the refrigerator

Questions

1. Is this technique as effective for separating amino acids as paper chromatography?
2. What do your results tell you about the nature of amino acids?
3. What results would you expect if you used a more acid and a more alkaline buffer?

121

Requirements

In addition to the basic equipment required for all electrophoresis work, the following will be required for this experiment:

Apparatus

Centrifuge or filter funnel and paper

Reagents

Pyridine   Acetone
Glacial acetic acid   Ninhydrin
Amino acids

Biological

A range of fruits

An investigation of dialysis

A mixed solution may contain both small and very large molecules. These molecules of different size may be separated by dialysis, which is in effect, diffusion through a selective membrane. The membrane most frequently used for small scale laboratory studies is Visking tubing (a specially prepared proprietary cellulose tube).

Dialysis is of considerable importance in biochemical work as it provides a convenient way of separating large enzyme molecules from inhibitors which may be present in an extract. Enzymes are proteins and may be separated from a mixture of enzymes and other proteins by carefully adjusting the isoelectric point and adding a strong solution of ammonium sulphate or ethanol. The precipitated protein may then be centrifuged off, redissolved and then purified by dialysis.

The rate at which dialysis occurs is affected by a number of factors, the most important of which are the pore size of the membrane and the amount of stirring in the water surrounding the outside of the membrane. The diffusion of electrolytes may be considerably speeded up if a difference in potential is established between the two sides of the membrane—the process is then called electrodialysis.

Procedure

1. Take a piece of Visking tubing about 20 cm long. Open out the tube—use distilled water to soften it and then roll it between the thumb and forefinger if necessary to separate the sides.
2. Tie a very tight knot in one end of the softened tubing.
3. Place about 20–30 ml of the solution provided in the dialysis

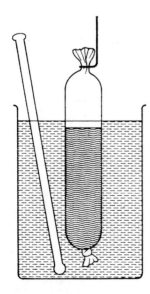

sac and tie a knot in the open end (if this is found to be difficult the open end may be closed by tying with a thread which can then be used to suspend the sac).

4. Place the sac in a large beaker containing at least 500 ml of distilled water.
5. Stir every few minutes for 1 hour.
6. Test a 2 ml sample of the water in the beaker by adding silver nitrate solution. (White precipitate indicates chloride ions.)
7. Take 2 ml of the water from the beaker and add 1 ml Millon's reagent. Heat. (Red coloration indicates protein.)

Requirements

Apparatus

Visking tubing (available from Phillip Harris Ltd)
1,000 ml beaker

Thread
Test tubes

Reagents

Solution containing casein and sodium chloride

Silver nitrate solution
Millon's reagent

The influence of temperature on enzyme activity

Enzymes are organic catalysts which influence the rate of one specific reaction. They permit the free energy of a system to be redistributed at body temperature so that biochemical reactions can occur at a fairly rapid rate. The enzymes are protein molecules and as such they are liable to progressive disruption and coagulation above certain critical temperatures. The critical temperature differs slightly from one enzyme to another. Below the critical temperature at which the enzyme is denatured there is an optimum temperature at which each enzyme catalyses its particular reaction at the maximum rate. Below the optimum temperature, the rate of an enzyme catalysed reaction obeys the usual laws of physical chemistry.

Procedure

Preparation of enzyme

1. Wash out the mouth with 100 ml distilled water to give a dilute solution of salivary amylase.
2. Add 60 ml M/5 disodium hydrogen phosphate solution and 40 ml M/5 potassium dihydrogen phosphate solution to the

123

dilute saliva. This will give a large quantity of dilute salivary amylase, buffered at pH 7·0. (If commercial buffer tablets are available, add 100 ml of prepared buffer to the diluted saliva.) The diluted saliva may be used without buffering if preferred, but this will add one more variable, albeit slight in this case, to the experimental situation.

Experiment

3. Place 2 ml of buffered saliva in a test tube and place in a water bath or beaker containing ice to equilibrate.
4. Place 2 ml of starch solution in a second tube and place in a water bath or beaker of ice to equilibrate.
5. When the tubes have had 5 minutes in the water bath to equilibrate, the buffered saliva is poured into the tube containing starch solution.
6. A sample of the reaction mixture is removed with a glass rod every 15 seconds and added to a spot of iodine solution on a white tile.
7. Record the time taken to reach the end-point, i.e. when no blue-black or reddish brown coloration occurs.
8. Repeat stages 3–7 at a series of temperatures as follows:
   In Ice. (Record reaction mixture temperature with a thermometer.)

   > Ambient or 20°C
   > 30°C
   > 40°C
   > 50°C
   > 60°C
   > 70°C

9. Plot a rough graph with the readings obtained (Temperature against Time in minutes). From this rough graph you should be able to see the temperature ranges in which the optimum temperature and the critical temperature lie. The rough graph will also enable you to select the most appropriate scale for each axis of your final graph.
10. Repeat stages 3–7 at several temperatures near the optimum and near the critical point. If time permits use steps of 2·5°C so that the parts of the graph where the rate of reaction is changing most can be plotted accurately.
11. Plot your final graph.

If a number of water baths are available they can be set at different temperatures. If only one bath is available start at 20°C and increase the temperature in 10°C steps, allowing plenty of time for equilibration.

N.B. The dial on a thermostatically controlled water bath is rarely accurate and the temperature of the reaction mixture should always be recorded with a thermometer. Do not waste time in the first run of experiments adjusting your bath precisely, as long as an accurate temperature is recorded with a thermometer for each reaction a good line can be plotted.

Requirements

Apparatus

Thermostatically controlled water bath(s)
Test tubes
Glass rod
White tile
2 ml pipette or 10 ml graduated pipette

250 ml flask or beaker
Measuring cylinder (if buffer is to be prepared by the student)
Stopclock or watch with sweep second hand
500 ml beaker (for ice)

Reagents

5% starch solution
pH 7·0 buffer solution of M/5 Na_2HPO_4 and M/5 KH_2PO_4

Solution of iodine in potassium iodide
Ice

Biological

Saliva (from student). Occasionally one meets an individual with an amylase deficiency in his saliva

---

The influence of pH on enzyme activity

Enzymes are organic catalysts which influence the rate of one specific type of reaction. They may be specific for breaking or making one specific kind of bond or even a single bond in a particular substance. The enzyme is a protein molecule which operates by becoming temporarily interlocked with the substrate molecule (template, or lock and key theory), and its ability to catalyse the reaction depends upon the closeness of this fit. The pH of the medium affects the character of the ionic binding both within the enzyme molecule, and between adjacent enzyme molecules. Any given enzyme molecule will, therefore, only 'fit' its substrate molecule over a limited pH range.

Procedure

1. Prepare a range of phosphate buffers, either from commercial buffer tablets or from solutions of M/5 anhydrous di-sodium hydrogen phosphate and M/5 potassium di-hydrogen phosphate as shown in the table overleaf.
2. Place 3 ml buffer in 12 test tubes to give a range from pH 5·3 to pH 8·05. Check the accuracy of your solutions with a pH meter if one is available.

pH	Na_2HPO_4	KH_2PO_4
5·3	2·5 ml	97·5 ml
5·6	5	95
5·9	10	90
6·2	20	80
6·5	30	70
6·65	40	60
6·8	50	50
7·0	60	40
7·15	70	30
7·4	80	20
7·75	90	10
8·05	95	5

3. Obtain a sample of saliva and dilute to 15 ml. Shake well.
4. Add 1 ml of diluted saliva to each tube.
5. If possible, use a thermostatically controlled water bath at 25°C. If a water bath is not available, the experiment may be carried out at room temperature. Fluctuations during the experiment will be minimal if the tubes are immersed in water which has assumed room temperature by standing in a large beaker for 30 mins.
6. Add 1 ml of starch solution to three of the tubes—do not attempt more as accurate readings will be impossible.
7. At intervals of 15 seconds, remove a sample of the reaction mixture and test it with iodine on a white tile. Record the time taken for the complete digestion of starch in each tube (iodine gives blue-black coloration with starch, reddish brown with dextrins and finally no colour when the starch has been completely digested to sugar).
8. Repeat with the remaining tubes in groups of three.
9. Plot a graph of your results—pH against time.
   What is the optimum pH for the action of salivary amylase?
   Is there a critical point above or below which no action occurs?

Requirements

Apparatus

Test tubes
Thermostatic water bath or large beaker
10 ml graduated pipette

Burette (for measurement of buffer solutions)
pH meter (optional)
White tile
Glass rod

Reagents

Solution of iodine in potassium iodide
Starch solution
M/5 di-sodium hydrogen phosphate (anhydrous)

M/5 potassium di-hydrogen phosphate
Store these stock solutions in a refrigerator

Biological

Saliva

126

The effect of substrate concentration on the reaction velocity of an enzyme catalysed reaction

The enzyme, kidney alkaline phosphatase, liberates phenol from phenyl phosphate. The free phenol can be estimated from the intensity of blue colour produced with Folin and Ciocalteus reagent.

Michaeli's constant can be determined from the results. This gives a measure of the affinity of the enzyme for the substrate and evidence for the formation of an enzyme-substrate complex.

Procedure

Preparation of enzyme

1. Homogenise a rat or mouse kidney in 10 ml of water. This can best be done by grinding it in a mortar with a pestle after first cutting it into slices with a scalpel.
2. Centrifuge the homogenate at 3,000 rev/min for 3 minutes (i.e. stop 2 on the Piccolo centrifuge).
3. Decant off the supernatant fluid and make up to 100 ml with de-ionised water.
4. Add a few drops of chloroform.

   This stock solution will retain its activity for at least one month if kept in a refrigerator at 4°C (near the ice box).

5. To prepare the enzyme from the stock solution, immediately prior to the experiment dilute it 1 part in 10 with de-ionised water. It may be necessary to alter this dilution to fit the readings on the colorimeter scale as the strength of the enzyme seems to differ quite considerably from animal to animal.

Buffer

To prepare this, add glycine, NaOH and NaCl together in a beaker observing the pH value of the solution with a pH meter. The optimal pH is 9·4. If a pH meter is not available, add 2 ml of 0·1 MNaOH to a solution consisting of 7·5 g of glycine + 5·85 g of NaCl in one litre. 8 ml of this latter solution must be added to the 2 ml of NaOH. This gives a solution with a pH of 9·22.

Substrate

Prepare a mM and a 10 mM solution of disodium phenyl phosphate. From this prepare tubes of stock substrate in the following way:

	Stock substrate	Water	Concentration obtained
	(ml)	(ml)	(0·01 mM)
mM	A 0·9	9·1	9
	B 3·0	7·0	30
10 mM	C 0·9	9·1	90
	D 3·0	7·0	300

Experiment

1. Set the following mixtures in test tubes:

Tube	Buffer (ml)	Substrate (ml)	Enzyme (ml)	Final substrate concentration
1	1	1 of A	1	3
2	1	1 of B	1	10
3	1	1 of C	1	30
4	1	1 of D	1	100
5	1	1 of water	1	—

2. Thoroughly mix the contents of each tube and incubate them in a water bath at 37°C for 10 minutes.
3. At the end of this time, add 2 ml of Folin and Ciocalteu's reagent to each tube. This stops the enzyme action and provides the blue colour. The Folin and Ciocalteu's reagent, which can be purchased, must be diluted, 1 part of reagent to 3 of de-ionised water.
4. Thoroughly mix the reagent with the contents of each tube.
5. Add 2 ml of MNa_2CO_3 to each tube, invert to mix and incubate for a further 10 minutes at 37°C.
6. With the red filter in place, adjust the needle of the colorimeter to zero first with de-ionised water and then with the blank solution, in tube 5.
   Read the intensity of blue colour in tubes 1 to 4.

Results

The colorimeter reading for each tube is proportional to the amount of phenol formed, and hence this is a measure of the reaction velocity of the phosphatase reaction.

Plot a graph of the substrate concentration against the reaction velocity.

From this graph determine the value of Michaeli's constant, which gives a measure of the affinity of the enzyme for the substrate. This constant is equal to that substrate concentration which gives half the maximum velocity. It should be expressed in terms of moles per litre.

Question

Is there a maximum substrate concentration, above which no further increase in reaction velocity can be obtained?

Requirements

Apparatus

Mortar and pestle
Vessel for killing a rat or mouse in
Centrifuge capable of running at 3,000 rev/min
Refrigerator

pH meter (optional)
Water bath at 37°C
100 ml measuring cylinder
10 ml graduated pipette
Colorimeter with red filter

Reagents

Chloroform
Glycine
NaOH

NaCl
Folin and Ciocalteu's reagent
M Na_2CO_3

Biological

Rat or mouse

The effect of enzyme concentration on the reaction velocity of an enzyme catalysed reaction

Procedure

Enzyme and Buffer

Prepare these as for the previous experiment.

Substrate

10 mM solution of disodium phenyl phosphate.

Experiment

1. Set up the following solutions in test tubes:

Tube	Buffer (ml)	Substrate (ml)	Water (ml)	Enzyme (ml)
1	1	1	0·8	0·2
2	1	1	0·6	0·4
3	1	1	0·4	0·6
4	1	1	0·2	0·8
5	1	—	1·5	0·5

129

2. Mix these tubes and incubate them at 37°C for 10 minutes.

3. At the end of this time, remove each tube from the water bath and add 2 ml of Folin and Ciocalteu's reagent to each tube. This reagent should be diluted one part in three with distilled water.

   This reagent stops the enzyme reaction and provides the reagent for the development of the blue colour.

4. Add 2 ml of M Na_2CO_3 to each tube and incubate again at 37°C for a further 10 minutes. Invert each tube to thoroughly mix the contents before incubating.

5. With the red filter in place, adjust the colorimeter needle to zero, first with de-ionised water and then with the blank solution, tube 5.

   Read the intensity of blue colour in each of tubes 1 to 4.

Results

The colorimeter reading for each tube is proportional to the amount of phenol formed. This is a measure of the speed of the enzyme catalysed reaction.

Plot a graph of reaction velocity against enzyme concentration.

Question

Is there a maximum enzyme concentration above which no further increase in reaction velocity can be obtained?

Requirements

Apparatus

Mortar and pestle
Vessel for killing rat or mouse
Centrifuge capable of running at 3,000 rev/min
Refrigerator

Water bath at 37°C
100 ml measuring cylinder
10 ml graduated pipette
Colorimeter with red filter

Reagents

$CHCl_3$
Glycine
NaOH

NaCl
Folin and Ciocalteu's reagent
M Na_2CO_3

Biological

Rat or mouse

Enzyme concentration and reaction velocity in enzyme catalysed reactions

To study the effect of varying the enzyme concentration on the reaction velocity of a biochemical reaction.

Procedure

1. Grind several 5 day old barley seedlings in a mortar with a little silver sand and 50 ml of de-ionised water.
2. Pour the amylase extract into a centrifuge tube and centrifuge for 5 minutes at 500 G.
3. To determine the period for which the experiment should be run mix together at zero time 5 ml of the extract and 5 ml of 0·4% soluble starch solution.
   Remove 3 drops from this mixture every 30 seconds and place on a spotting tile.
   Test with a drop of iodine in potassium iodide solution.
   Note the time of the first appearance of a reddish colour.
   Use half of this time as the experimental period.
4. Prepare the enzyme solution in the following strengths: 100%, 75%, 50% and 25%.
5. Place 10 ml of each of these solutions in a separate beaker.
6. Add 10 ml of 0·4% soluble starch solution to each of these beakers in turn.
7. Mix thoroughly.
8. At the end of the experimental period, pipette 5 ml of the experimental mixture into a 100 ml volumetric flask containing 5 ml of iodine in potassium iodide solution. This immediately stops the reaction.
9. Add de-ionised water to bring the volume up to 100 ml.
10. Pour the solution into a conical flask and set aside until all the experiments have been run.
11. Clean the volumetric flask and repeat the procedure with the next enzyme strength.
12. When all the strengths have been run, prepare a zero enzyme concentration flask by adding 10 ml of 0·4% soluble starch solution to 10 ml of de-ionised water.
13. Mix thoroughly.
14. Pipette 5 ml of this solution into 5 ml of the iodine in potassium iodide solution in a 100 ml volumetric flask.
    Make up to 100 ml with de-ionised water.
15. The amount of reaction which has taken place in each of these experiments will be measured with a colorimeter.
    First the colorimeter must be set to zero. To do this use a solution of 5 ml of iodine in potassium iodide made up to 100 ml with de-ionised water.
    Determine the light absorption due to the starch-iodine colour complex in each of the five tubes.

Results

Consider the absorption of the control to be 100, and determine the other readings as percentages of this.

Plot a graph of these percentages against the strength of the enzyme.

131

Requirements

Apparatus

Mortar and pestle
Centrifuge
Spotting tile
Graduated 10 ml pipette
4 beakers
100 ml volumetric flask
4 conical flasks
Colorimeter

Reagents

Silver sand
0·4% soluble starch
Iodine in potassium iodide

Biological

Barley seedlings which are about
5 days old

An experiment to demonstrate the inhibition of an enzyme reaction

An enzyme operates by becoming temporarily interlocked with the substrate molecule (template or lock and key theory). Certain chemicals known as inhibitors will stop an enzyme action when they are added to the system. Inhibitors usually have a small molecule and can be separated from the enzyme by dialysis. In research on enzyme activity, the inhibition of a reaction by a particular chemical may be used to pinpoint the sites which are concerned with the stereochemical fit. For the purposes of this demonstration a reaction which can be observed visually has been chosen.

Procedure

1. Remove a small piece of liver from a freshly killed mammal.
2. Place the liver in 5 ml hydrogen peroxide in a boiling tube.
3. Streams of oxygen bubbles are released at the cut surface of the liver as the liver peroxidase splits the hydrogen peroxide into water and oxygen.
4. Add a few drops of potassium cyanide solution (or 1 very small crystal of potassium cyanide). Note that the effervescence ceases almost immediately.

WARNING: Potassium cyanide is extremely poisonous and should be handled with great care.

5. If desired the mixture may be dialysed with a number of changes of water to remove the inhibitor.

Wash out all apparatus which has contained cyanide with copious supplies of water immediately after use. In a biological laboratory there is a risk that contaminated glassware may be used with living material later.

Requirements

Apparatus

Test tube
Dissection instruments

Materials for dialysis (if required)

Reagents

Potassium cyanide

Biological

Freshly killed mouse or rat

The use of an enzyme inhibitor to sort out the individual steps in the pathway of a reaction

In addition to using an inhibitor to sort out the pathway of a chemical reaction, this experiment also shows the function of a dehydrogenase in the tricarboxylic acid cycle.

The principle behind the use of inhibitors for studying reaction pathways is as follows. Imagine a reaction sequence

$$A \rightarrow B \rightarrow C \rightarrow D \rightarrow E$$

in which we believe $C \rightarrow D$ to be controlled by a particular enzyme. If we know an inhibitor for the $C \rightarrow D$ part of the reaction then we compare the existence of these substances before and after inhibition. The inhibited reaction should show much the same amount of A and B as the normal reaction but there should be no D or E formed. To confirm this add D to the inhibited reaction and observe that E is now formed.

The example chosen to illustrate this is the conversion of succinic acid to fumaric acid as in the tricarboxylic acid cycle. This is inhibited by malonic acid.

The activity of the succinic dehydrogenase could be measured determining the rate of uptake of oxygen with added succinate. But the rate of electron transfer is usually slow when it is via the cytochromes to oxygen as compared with the potential rate removal of electrons from succinate.

Usually it is quicker to measure the change in colour of a dye which is reduced as a result of the activity of the dehydrogenase. This dye must not react significantly with oxygen. Such a dye is 2,6-dichlorophenol indophenol (DCPIP). Dilute alkaline methylene blue can be used also. The change in colour of the dye is more rapid if no air is present. This can either be achieved by keeping the reaction mixtures under a layer of paraffin or performing the experiments in Thunberg tubes or adding N methyl phenazinium methosulphate (phenazine methosulphate or PMS for short).

This increases the rate of electron transfer from the enzyme to the dye.

Procedure

Preparation of suspensions

1. Remove the seed coat from 10 germinating bean seeds.
   Grind the seeds in 20 ml of ice-cold 0·4 M sucrose solution until the preparation is like a pulp.
   Centrifuge the suspension for 5 minutes at 500G to remove the cell debris.
   Use the supernatant as a source of enzymes.
2. Remove a kidney from a freshly killed rat or mouse.
   Rinse it under the tap and dry on a piece of filter paper.
   Grind it in a mortar in about half its volume of silver sand.
   Add 5 ml of de-ionised water and mix well by grinding.
   Pour the mixture into a test tube and further mix by shaking vigorously for 10 seconds.
   Immediately before use mix well by inverting several times.
3. Set up the following test tubes:
   Tube 1. 2 ml phosphate buffer, 1 ml of dilute methylene blue, 1 ml sodium succinate (0·1 M), 1 ml of suspension.
   Tube 2. as for tube 1 but using boiled suspension.
   Tube 3. 1 ml of phosphate buffer, 1 ml of methylene blue, 1 ml of succinate, 1 ml of suspension, and 1 ml of sodium malonate (0·1 M).
4. Mix the contents of each tube thoroughly and add sufficient liquid paraffin to prevent oxygen entering the tube.
5. Then incubate at 37°C for 20 minutes, and note the colour changes at the end of this time.
   Alternatively the colour change could be followed with a colorimeter. In this case use 10 ml colorimeter tubes instead of test tubes. Each tube should be corked with a 1-hole stopper arranged with a short piece of glass tubing in it to which is attached a piece of rubber tubing.
   Zero the colorimeter with de-ionised water.
   Evacuate each tube for 1 minute while continually shaking the contents.
   Immediately after evacuating each tube, make a colorimeter reading. Note the exact time and begin the incubation period.
   Take a reading on each tube every 5 minutes.

Note. When using plant material the pH of the phosphate buffer should be 6·8, and when using kidney the pH should be 7·4.
If DCPIP is used, use 0·1 ml of a mM solution.
As already mentioned Thunberg tubes could have been used and PMS could also have been used. Use 0·2 ml of 20 mM PMS.

Results

Plot the colorimeter readings against time.

Question

When using paraffin, what would happen to tube 1 if at the end of the incubation period, it was shaken and then incubated again?

Requirements

Apparatus

Mortar and pestle
Centrifuge
4 test tubes
Incubator set at 37°C
Colorimeter tubes set up as shown in the diagram
3 Thunberg tubes
Filter pump

Reagents

ice-cold 0·4 M sucrose
Silver sand
Phosphate of 6·8 pH for plant material or 7·4 for kidney
Dilute methylene blue
0·1 M sodium succinate
0·1 M sodium malonate
Liquid paraffin
mM DCPIP
20 mM PMS

The synthesis of starch

There are two classes of enzymes which hydrolyse starch, namely amylases and phosphorylases. The former take part in a reaction which is irreversible, while the latter take part in a reversible reaction. Amylases produce maltose units which require maltase to complete the conversion to glucose. Phosphorylases produce glucose-phosphate units. Phosphatase breaks these units apart.

This experiment attempts to show the reversibility of the phosphorylase reaction.

Procedure

Preparation of the enzyme. Make up fresh just before the experiment.

1. Peel a potato and cut it into pieces.
2. Place it in a blender with 40 ml of 0·01 N potassium cyanide. The cyanide is to inhibit the activity of the phosphatase. Grind for 45 seconds.

WARNING: Potassium cyanide is extremely poisonous and should be handled with great care. Do not use a mouth pipette.

3. Filter the homogenate through muslin or through a Buchner funnel.
4. Centrifuge the filtrate for 5 minutes and decant off the supernatant.
5. A portion of this should give a negative test with iodine.
6. The supernatant contains the enzyme.

135

Wash out all apparatus which has contained cyanide with copious supplies of water immediately after use.

Experiment

1. Set a series of tubes as follows:
   Tube 1. 3 ml of 0·01 M glucose, 1 drop of 0·2% starch solution.
   Tube 2. 3 ml of 0·01 M glucose-1-phosphate, 1 drop of 0·2% starch solution.
   Tube 3. 3 ml of 0·01 M glucose-1-phosphate.
   Tube 4. As for tube 2.
   Tube 5. 3 ml of 0·01 M glucose-1-phosphate, 1 ml of 0·2 M KH_2PO_4, 1 drop of 0·2% starch solution.
   Tube 6. 1 ml of 0·2 M KH_2PO_4, 3 ml of 0·2% starch solution.
   Tube 7. As for tube 6.
2. To tubes 1, 2, 3, 5 and 6, add 3 ml of the enzyme extract.
   To tubes 4 and 7, add 3 ml of the boiled enzyme extract.
3. Immediately the enzyme is added, mix the liquids thoroughly, and spot a few drops from each tube on to drops of I_2KI solution on a spotting tile.
4. Repeat this test every 3 minutes until the reaction is completed (usually about 30 minutes).

Questions

1. What is the function of the KH_2PO_4?
2. In preparing the enzyme, why is it important to inhibit the action of the phosphatase with potassium cyanide?
3. In this experiment, five reagents are used. In writing up your results you should make clear the function of each of these reagents, and thereby list the requirements for the enzymatic synthesis of starch with phosphorylase.

Requirements

Apparatus

Blender
Buchner funnel and filter pump or muslin
7 test tubes

Beaker
Centrifuge
Spotting tile
Glass rods

Reagents

40 ml of 0·01 N potassium cyanide
I_2KI
3 ml of 0·01 M glucose

0·2% starch solution
12 ml of 0·01 M glucose-1-phosphate
3 ml of KH_2PO_4

Biological

Potato

An analysis of the molecules present in starch

Starch is a mixture of two types of molecules, amylose and amylopectin.

Amylose with 1-4 α links between the glucose molecules

Amylopectin with 1-6 α links between the glucose molecules

Starch can be depolymerised by hydrolysis of the glycoside bonds by acids or enzymes known as amylases. The amylases produce only partial hydrolysis as they can not attack the 1-6 bonds.

There are two types of amylases, α and β; α-amylases such as the salivary and pancreatic ones, attack bonds at random in the interior of the chain to produce oligosaccharides which are slowly split to maltose; β-amylases attack the bonds at the non-reducing ends of the chain only producing maltose units. β-amylases are found in plants and micro-organisms.

With both these enzymes hydrolysis stops at the points branching. For further hydrolysis to take place, glucosidases must be present to split the 1-6 bonds.

Starch in water produces a very viscous medium. Hydrolysis reduces this viscosity. Straight chain polymers show a greater viscosity than ones which are branched.

Hydrolysis also increases the number of accessible reducing groups.

Procedure

Preparation of the enzymes

1. α-amylase.
   Wash out the mouth with 100 ml of de-ionised water to give a dilute solution of the enzyme.

137

2. β-amylase.

Soak about 50 barley seeds and allow them to germinate on moist filter paper for 4–5 days at room temperature. Then grind up the seedlings in a mortar with silver sand and de-ionised water to a fine mush. Filter off the liquid which will contain the enzyme.

Changes in viscosity as a measure of the rate of hydrolysis

1. Attach a piece of capillary tubing of $\frac{1}{4}$ mm internal diameter to the tip of a graduated 5 ml pipette with a short piece of rubber tubing.

   Measure the viscosity of the liquid by measuring the time taken for 1 ml of liquid to pass between the 3 to 4 ml marks. This viscometer should deliver 1 ml of water in 20–30 seconds when held in a vertical position.

2. Pour 10 ml of 2% starch solution into each of 4 test tubes.
   If pipetting, take care not to add any saliva as this will ruin the experiment.

3. Heat 2 ml of each enzyme in a boiling water bath for 10 minutes to inactivate it.

   Add 1 ml of this inactivated enzyme to each of two of the tubes. These will then serve as controls.

   Do not add active unheated enzyme until you are about to begin the experiment.

4. Use the β-amylase first as small errors in timing are not as serious as with the α-amylase.

   Add 1 ml of β-amylase to one of the experimental tubes.

   Shake to mix the liquids quickly.

   Take the time of mixing as zero time, and start the stop watch.

   Immediately place the viscometer in the reaction mixture.

   Draw the liquid into the pipette until it is well above the starting mark.

   Keep the liquid above the starting mark by holding a finger on the top of the pipette until $1\frac{1}{2}$ minutes have elapsed from zero time.

   Now remove the finger from the pipette.

   When the liquid level reaches the 3 ml mark start a second stop watch.

   Stop the watch when the level passes the 4 ml mark.

   Repeat 2 minutes from zero time, and then at 5 minute intervals for 30 minutes and again at the end of an hour.

   While this experiment is running remove 8 drops of the reaction mixture at 15, 30, 45 and 60 minutes. Set these tubes, each containing 8 drops of the reaction mixture aside for estimation of the reducing activity.

   Start the α-amylase run after you have obtained the 30 minute reading with β-amylase.

   Use a clean pipette.

Estimation of reducing activity

1. Add 5 ml of Benedict's reagent to each of the samples prepared in the previous part of the experiment.
   Heat the tubes in a boiling water bath for 3 minutes.
2. Compare the colours obtained with a set of standards prepared thus: Add 5 ml of Benedict's reagent to each of 6 test tubes. To one of these add 8 drops of water and the others add respectively, 0·05%, 0·10%, 0·25%, 0·50% and 1·0% solutions of maltose to each of the other tubes.
   Heat these tubes over a boiling water bath for 3 minutes.
3. Estimate the amount of reducing sugar in the experimental tubes by comparing the amount of precipitate with the amount in the tubes of known quantities of maltose.

Analysis of the hydrolysates

1. Draw a line across a 25 cm² piece of filter paper 3 cm from one end.
   Along this line mark 5 evenly spaced pencil dots to mark the point of application of each of the samples.
2. The following solutions should be applied to the paper:
   (a) 4% glucose
   (b) 4% maltose
   (c) 4% glucose + 4% maltose
   (d) α-amylase hydrolysate of starch
   (e) β-amylase hydrolysate of starch
3. Place the paper in a chromatography tank using 50 ml of a solvent made up of 35 ml of n-butanol, 5 ml of ethyl acetate, and 10 ml of water.
4. Run overnight.
5. Remove the paper and dry thoroughly, after marking the solvent front.
6. Develop the spots in a reagent made up of: 0·5 g of m-phenylene diamine, 1·2 g of stannous chloride, 20 ml of acetic acid and 80 ml of ethyl alcohol.
7. If the chromatogram has been allowed to run overnight, the solvent front will almost certainly have reached the top of the paper. In this case an R_g value should be calculated instead of the more usual R_f value. The R_g value is based on the distance travelled by the glucose as the reference point.

$$R_g = \frac{\text{distance the substance has run from the origin} \times 100}{\text{distance the glucose has run from the origin}}$$

   The R_g value for glucose itself will be 100.
8. Also note the colour of each spot.

Requirements

Apparatus

Mortar and pestle
Filter funnel and paper

2 lengths of capillary tubing of 0·25 mm internal diameter

2 graduated 5 ml pipettes
2 small pieces of rubber tubing to join the capillary tubing to the pipettes
Retort stand or burette stand
12 test tubes
Boiling water bath

1 ml graduate pipette
2 stop watches
Chromatography tank
25 cm² filter paper
Developing tray
Ruler

Reagents

De-ionised water
Silver sand
2% starch solution
Benedict's reagent
Maltose of the following strengths:
0·05%, 0·10%, 0·25%, 0·50%, 1·0%, 4%

Glucose, 4%
m-phenylene-diamine
Stannous chloride
Acetic acid
Ethyl alcohol
n-butanol
Ethyl acetate

Biological

Barley seedlings
Salivary amylase

Investigation of the chemical constitution of plants

The purpose of this section is to study the chemical constitution of plant tissues on a microscopic scale. It is assumed that studies on a macroscopic scale have been carried out as part of a study of human physiology or food chains at an earlier stage in the school.

Procedure

The investigations should be carried out in two stages firstly the demonstration of the test on a pure sample where the substance is known to be concentrated (e.g. starch in potato tubers) and, secondly the detection of the substances in tissues from various organs.

The tests should be carried out on fresh sections of various thicknesses. To localise chemicals within a cell accurately it must be very thin, but this means that there will be very little chemical present. Therefore the test may not be sufficiently sensitive.

When water-soluble chemicals are being investigated the section should first be blotted with filter paper. This will decrease the leakage from damaged cells.

For speed of operation several sections should be placed on a slide under one long coverslip.

Care should be taken to wipe the slide dry before placing it on the stage. Reagents should not be added to the slide while it is on the stage.

Carbohydrate tests

1. *Reducing and non-reducing sugars*

Prepare three sections and treat them as follows:

(a) Mount in Benedict's reagent.
(b) Mount in Benedict's reagent and boil for 3 minutes.
(c) Boil in 10% HCl in a watch-glass for 1 minute and neutralise with solid Na_2CO_3. Then treat as for (b).

Compare the three sections. (In (c) the reaction is partly due to the hydrolysis of hemicelluloses.)

2. *Sugars*

Immediately before use mix 1 part of phenylhydrazine hydrochloride in glycerol (1 : 10) with 2 parts of sodium acetate in glycerol (1 : 10).

Mount a thick section in 1 drop of the mixture, heat at about 100°C for 15 minutes. Then examine under the microscope.

Carry out these tests on pure sugars also to facilitate their recognition in the plant tissues.

3. *Pentoses*

Mount sections in a mixture of phloroglucinol (saturated in 95% alcohol) plus an equal volume of concentrated HCl. Observe the appearance of a purplish-red colour, due to lignin, then heat for 10 minutes and observe again. (Pentoses may occur as free sugars but usually result from the hydrolysis of polymers such as Xylan or Araban, which are constituents of the wall hemicelluloses, pectic materials and gums.)

4. *Inulin*

Sphaero-crystals appear when stored in 70% alcohol for a day or longer. The structure of these is best seen when mounted in chloral-hydrate iodine.

For confirmation of this mount the alcohol treated sections in 15% thymol in absolute alcohol, to which has added a drop of concentrated sulphuric acid.

5. *Starch*

Observe grains from different sources. For the starch-iodine test a weak solution should be used to prevent the blue colour masking the structural features of the cells.

6. *Cellulose*

Mount in iodine as above and note the position of any starch or other blue-staining substances. Blot off the excess iodine and replace by 75% H_2SO_4. Then examine under the microscope.

or, soak the sections for 15 minutes in 10% aq. congo red and then mount in 10% HCl (lignified walls remain unstained).

7. *Ligno-cellulose*

Place the section in a watch glass of alcoholic phloroglucin. When well soaked dip in concentrated HCl. Then examine under the microscope.

8. *Callose*

Stain for 20 minutes in 0·1% aq. aniline blue. Wash in water and mount in glycerol.

9. *Mucilage*

Mount the sections in basic lead acetate.

Protein tests

Amino groups

Mount the sections in 0·3% ninhydrin in a mixture of 19 parts butanol and 1 part 2 N acetic acid. Warm until nearly dry, and then mount in water.

or, treat the section on a slide with potassium ferrocyanide in acetic acid for 1 hour. Then rinse in 60% alcohol and treat with ferric chloride. Examine under the microscope.

Free nitrate ions

Immerse thick sections of the tissue in diphenylamine sulphate and examine under the microscope.

Free ammonium ions

These are sometimes found in the sap. Mince the tissue and strain off some of the sap through muslin. Add a few drops of 10% KOH to the sap in a test tube and warm gently. Test for ammonia by smell and by holding the stopper of the concentrated HCl bottle at the mouth of the test tube.

Lipid tests

1. *Fats, waxes and oils*

To a saturated solution of Sudan IV in absolute alcohol add water to give a 70% alcoholic solution. Stain the sections in this solution in a watch-glass for 5 minutes. Then wash rapidly in absolute alcohol and mount in 25% glycerol.

2. *Cutin and suberin*

Mount in chromic acid.
To distinguish cutin from suberin heat in 30% KOH and then examine.

Tests for phenolic substances

1. *Lignin*

As for ligno-cellulose or, mount the section in aniline sulphate.

2. *Tannin*

Mount in 10% $FeCl_3$ in methyl alcohol. This is a useful general test but it is not specific.

or, for a specific test, mount the sections in a mixture of 1 g sodium tungstate, 2 g sodium acetate and 10 ml of water.

Terpene tests

1. *Carotene*

The sections should be placed in corked specimen tubes containing a mixture of aq. KOH, 40% alcohol, and water (1 : 2 : 3) for 1–2 hours. Wash in water and place on a slide. Blot with filter paper and mount in concentrated H_2SO_4.

2. *Xanthophyll*

Place the sections on the slide in a drop of chloroform, followed a few minutes later by a drop of petroleum ether.

Mineral tests

Calcium carbonate, calcium oxalate, and silica can be separated by the following procedure: To a section on a slide add a drop of 10% acetic acid and observe. To a second section add a drop of concentrated sulphuric and observe.

Results

Carbohydrate tests

1. Benedict's reagent Reducing sugars give minute black grains of cuprous oxide	**2. Phenylhydrazine test** Fructose gives clusters of fine yellow needles (fructose osazone) Glucose is precipitated after $\frac{1}{2}$ hour Maltose gives clusters of short broad yellow crystals which are soluble in hot water (maltosozone)
3. Phloroglucinol test Xylose and arabinose give bright red colour	**4. Thymol test** Inulin crystals dissolve and give an orange colour. Re-precipitation of crystals which darken to purple
5. Iodine/KI test Starch gives a blue colour which disappears on heating to 70°C	**6. I_2 in KI/H_2SO_4 test** Cellulose swells and becomes bright blue or, Congo red, cellulose gives blue
7. Phloroglucinol test Ligno-cellulose gives a bright red colour	**8. Aniline blue test** Callose gives a bright blue colour
9. Lead acetate test Mucilages give a granular yellow precipitate	

Protein tests

1. Ninhydrin test
Amino groups give a purple colour

2. Potassium ferrocyanide test
Proteins give a blue colour

Test for free nitrate ions

Diphenylamine colours them blue

Lipid tests

1. Sudan IV test
Fats, waxes and oils selectively take up the red stain

2. Chromic acid/KOH test
Cutin and suberin dissolve in the acid while the addition of 30% KOH deepens the colour of suberin which exudes drops of oil

Tests for phenolic substances

1. As for ligno-cellulose

2. Ferric chloride test
Tannin gives a blue-black colour
Tungstate test gives yellow-brown colour

Terpene tests

1. KOH/alcohol/water test
Orange deposits of carotene slowly replaced by dark blue microscopic crystals

2. CHCl₃/Petroleum ether test
Xanthophyll gives yellow crystals

Mineral tests

	$CaCO_3$	Calcium oxalate	Silica
10% acetic acid	Solution, with effervescence	Insoluble	Insoluble
Conc. H_2SO_4	Solution, with effervescence and precipitation of fine crystals of $CaSO_4$	Solution without effervescence and with precipitation of fine crystals of $CaSO_4$	Insoluble

Requirements

Reagents

Benedict's reagent
HCl, 10% and concentrated
H_2SO_4, 75% and concentrated
Acetic acid, 10% and 2N
KOH, 10% and 30%
Solid Na_2CO_3
Chromic acid

Sodium acetate
Ferric chloride
Potassium ferrocyanide
Alcohol, 40%, 60%, 70% and 95%
Butanol
Methyl alcohol
Sodium tungstate

Phenylhydrazine hydrochloride
Glycerol, 25% and pure
Phloroglucinol
Weak iodine/potassium iodide solution
(3 g I_2 plus 15 g KI per litre of water)

0·1% aniline blue
0·3% ninhydrin
15% thymol
Diphenylamine sulphate
Chloroform
Petroleum ether

Biological materials

Reducing sugars—leaf sections cleared of chlorophyll, or sections of the fleshy leaves of onion bulbs

Inulin—*Dahlia* root tubers

Starch—Potato stem tubers and leaf sections

Cellulose—*Helianthus* or any plant with collenchymatous tissue

Mucilage—stem sections of *Tilia* (recognised by large size, disorganised appearance and jacket of starch cells around them)

Fats and oils—leaves of *Dactylis glomerata*, *Mentha aquatica*, endosperm of *Helianthus annuus*

Waxes—leaves of conifers, stems of cacti (*Cereus* spp.)

Cutin—leaves of *Rhododendron* and cherry laurel

Suberin—outer layers of old stems and roots

Lignin—woody tissues, xylem and sclerenchyma

Carotene—carrot roots (crystal-like bodies associated with starch grains)

PHYSIOLOGY

The investigation of a plant's mineral requirements

The purpose of this investigation is to observe the effects of omitting a single element from the plant's complete nutrient medium. Several elements are examined and a control experiment is performed.

Procedure

1. Fill the bottles with the required nutrient.
2. Cover them with black paper to reduce the growth of algae.
3. Label each bottle.
4. Place a seedling in each so that the seed is just below the top of the cork. Hold it in place with cotton wool. This must be dry and should be replaced when damp to reduce the risk of fungal infection.
5. Any losses due to transpiration and evaporation should be made good by topping up with the appropriate culture solution.
6. The bottles should be kept in a well-lit place.
7. Aerate daily by pumping air through the tube provided. This is especially important in the early stages when the water level must be high. Otherwise the plants will drown. When the roots are longer an air pocket should be left at the top of the solution.
8. Regularly check the pH of the solutions. To do this slightly lift the cork and plant, and holding a test paper in a pair of forceps, dip it in the solution. Most plants prefer a slightly acid medium, pH 5 to 6.

   Minute amounts of extremely dilute sulphuric acid or potassium hydroxide may be added to adjust the pH.
9. It is good to dip the roots in de-ionised water occasionally to remove any chemical deposits.

Results

Compare the plants grown on the deficient solutions with the one on the complete medium.

Results could be noted under the following headings:
growth, leaf colour, leaf formation, root system, and differences between young and old leaves.

Requirements

Apparatus

Third pint milk bottle
Two-hole cork to fit the bottle. Glass tube in one hole for aeration purposes
The other hole should be enlarged to fit the plant

Sterilise in an autoclave if available, otherwise wash in boiling water and then in de-ionised water
Black paper for each bottle

Reagents

The complete medium is made up as follows: (after Jamel, *Introduction to Plant Physiology*, Clarendon Press, Oxford)

$CaSO_4,2H_2O$	0·25 g
$Ca(H_2PO_4)_2,H_2O$	0·25 g
$MgSO_4,7H_2O$	0·25 g
NaCl	0·08 g
KNO_3	0·70 g
$FeCl_3,6H_2O$	0·005 g

Make up to one litre with de-ionised water

For cultures lacking various elements substitute as follows:

Potassium	replace	KNO_3	by 0·59g $NaNO_3$
Calcium	replace	$CaSO_4, 2H_2O$	by 0·20g K_2SO_4
		$Ca(H_2PO_4)_2, H_2O$	by 0·71 g $Na_2HPO_4, I2H_2O$
Iron	omit	$FeCl_3, 6H_2O$	
Nitrogen	replace	KNO_3	by 0·52 g KCl
Phosphorus	replace	$Ca(H_2PO_4)_2, H_2O$	by 0·16 g $CaNO_{32}$
Sulphur	replace	$CaSO_4, 2H_2O$	by 0·16 g $CaCl_2$
		$MgSO_4, 7H_2O$	by 0·21 g $MgCl_2$

Biological

A series of small holes should be drilled in a thin sheet of cork. Float the cork in a trough of water. Place a seed over each hole. After seven to fourteen days the plants should be well developed and can be transferred to the culture bottles. Alternatively some seeds will germinate in sand moistened with the culture medium

---

The investigation of mineral deficiency in ' Aspergillus niger '

Procedure

1. Prepare a culture of *A. niger*.
2. Transfer 3–4 loopfuls of spores from this culture aseptically to a 0·9% sterile sodium chloride solution.
3. Shake the suspension to disperse the spores as evenly as possible.
4. Prepare nine 125 ml conical flasks, each containing 20 ml of medium.
   One flask should contain the complete growth medium, while each of the other flasks should be deficient in one element.
   Stopper each flask with cotton wool.
   Autoclave each flask.
5. Inoculate each flask with three loopfuls of *A. niger* suspension.
6. Incubate at 25°C or room temperature for a week.
   During this time observe the time at which spores first appear and the colour of the growth.

7. At the end of the week, harvest each of the growths by lifting it off the medium with a spatula, roll the growth into a ball and, on a piece of filter paper, gently squeeze as much of the liquid out as possible with your fingers.
8. Weigh each of the harvested mats.

Results

Which minerals are required by *A. niger*?
Note any effects mineral deficiency has on the time of spore production and the colour or form of the growth.

Requirements

Apparatus

Microbiological loop
9 125 ml conical flasks
Spatula

Reagents

Aspergillus complete medium
Prepare the mineral solutions from **Analar reagents** thus:
0·05 g of sodium versenate in 10 ml of de-ionised water (disodium ethylenediamine tetra-acetic acid)
0·02 g of K_2HPO_4 in 1 ml of de-ionised water
0·08 g of $MgSO_4.7H_2O$ in 1 ml of water
0·2 g of NH_4Cl in 10 ml of water
0·05 g of $ZnCl_2$ in 10 ml of water
0·008 g of $FeSO_4.7H_2O$ in 1 ml of water
0·006 g of $NaMoO_4.2H_2O$ in 1 ml of water
0·004 g of $CuSO_4.5H_2O$ in 1 ml of water
0·004 g of $CoSO_4.7H_2O$ in 1 ml of water

0·005 g of $MnSO_4.H_2O$
These reagents should be added to 500 ml of de-ionised water
Mix carefully after the addition of each reagent and then make the total volume up to 1 litre
Add 2 g of glucose to this solution

Deficient medium

These media can be prepared by adding the amounts indicated above except for the compound which the medium is to be deficient in
Otherwise prepare these media as for the complete medium
Ammonium chloride and glucose must be added to all media
Check the pH value of all media before pouring into the 125 ml flasks. With dilute KOH, adjust the value to 6·5–6·9
After each flask has been stoppered with cotton wool, it must be autoclaved

Biological

Aspergillus niger

The permeability of the cell membrane

To show that the membrane is selective in only permitting certain substances to enter the cell.

Procedure

1. Pour a drop of 0·02% neutral red solution into a test tube and add 1% sodium bicarbonate, a drop at a time until the colour changes.
2. Add dilute acid, a drop at a time until the colour changes again.
3. Note these colour changes.
4. Add 25 ml of 0·02% neutral red to 25 ml of yeast suspension in a flask. Observe the colour and note any colour changes for five minutes.
   (Neutral red can enter cells without killing them and it goes through a colour change at about pH 7·2.)
5. Centrifuge 10 ml of the suspension and note the colours of the cells and of the liquid.
6. Boil 10 ml of the original coloured suspension for a few minutes over a water bath. Note any colour changes.
7. Add 1 ml of 0·01 M sodium hydroxide to 10 ml of the original coloured suspension and note the resulting colour.
8. Add 1 ml of 0·01 M potassium hydroxide to 10 ml of the original suspension and note the resulting colour.
9. Add 1 ml of 0·01 M ammonia solution to 10 ml of the original suspension and note the resulting colour.

Questions

1. Explain the change in colour of the suspension after it had been in the stain for five minutes.
2. Explain the colour changes when the suspension is boiled.
3. Why does only one of the three bases used cause the suspension to change colour?
4. Is there any evidence for an active transport mechanism?

Requirements

Apparatus

Flask
Small measuring cylinder
Water bath. Teat pipettes
Centrifuge

Biological

Add an ounce of dried yeast to 250 ml of 1% NaHCO$_3$

Reagents

1% NaHCO$_3$
0·02% neutral red
0·05 M HCl
0·01 M NaOH (0·4 g/1)
0·01 M KOH (0·56 g/1)
0·01 M NH$_4$OH (10 ml of the concentrated solution added to 82 ml of water. Then 10 ml of this up to a litre)

The effect of salt concentration on water absorption and plant growth

Plant roots may be immersed in water and yet unable to take up much water because of an unfavourable diffusion pressure deficit between the external solution and the internal solution of the roots.

This condition is known as physiological drought.

Procedure

Place 200 ml of 0·01 M $CaCl_2$ in a flask.

Place 200 ml of 0·02 M, 0·03 M, 0·05 M and 0·1 M and 0·2 M $CaCl_2$ solutions in similar flasks.

In another flask, place 200 ml of distilled water.

Select seven uniformly vigorous seedlings of any easily grown plant, and support each in a hole in a cork sheet with cotton wool.

Place the cork over the flask so that the roots are in the solution.

Measure the length of the stem of each seedling above the cotyledons.

Mark the level of the solution in the flask.

At daily intervals top up the solution to this mark, noting the amount of solution required to restore the level.

After a week remove the plants and measure the length of the shoots above the cotyledons.

Observe the conditions of the plants.

Results

Set out the results under the following headings:

Strength of solution	Length of shoot above cotyledons	Total amount of solution used	Condition of plants

Requirements

Apparatus

7 250 ml flasks
7 small sheets of cork, each with a hole in the centre through which the plant can be placed
Cotton wool
Ruler

Reagents

300 ml of $CaCl_2$ solutions of the following strengths: 0·01 M, 0·02 M, 0·03 M, 0·05 M, 0·1 M and 0·2 M
Distilled water

153

An experiment to demonstrate active transport in animal tissue

Any movement of substances across a cell membrane against a concentration gradient will require a source of energy. Any movement requiring such work on the part of the cell is known as active transport.

Procedure

1. Remove the skin from the lower leg of a pithed frog by cutting through it just above the knee, going completely round the leg. Grasp the lower cut edge, loosen it from the connective tissue around the knee, and peel it off the leg as if it were a stocking.
2. Holding the skin as it is on the frog's leg; ligature the lower end.
3. Loop another thread around the knee end of the skin but do not tie completely. Insert a pipette containing Ringer's solution into the bag and allow the solution to flow in until the bag is full. Note the amount of fluid added.
4. Ligature the top of the skin tightly.
5. Immerse the bag in a beaker of Ringer's solution and aerate the solution by stirring.
6. Allow one hour for equilibration and then lift the bag out by one of the cotton ends, remove excess liquid by rubbing it against the lip of the beaker and place it in a covered weighing bottle.
   It must not be kept in the air too long as evaporation will take place.
   Weigh.
7. Return to the Ringer's solution.
8. Continue aerating frequently and weighing at 15 minute intervals.
9. When the reading has reached a peak, place the bag in a beaker of 0·002 M potassium cyanide.

WARNING: Potassium cyanide is extremely toxic. Mouth pipettes must not be used.

(KCN is a respiratory inhibitor and hence will inhibit active uptake.)

10. Continue weighing every 15 minutes for 1–1½ hours.
11. At the end of the experiment open the bag and cut away all the skin beyond the ligatures.
12. Blot on filter paper and lay it on a piece of graph paper. Trace around the edges and use the tracing to measure the area of the skin.

Results

Plot a graph of weight against time in minutes. Indicate on the graph the point at which the bag was placed in cyanide.

154

Determine the rate of active transport as mg water/cm²/hour.

Requirements

Apparatus

Graduated pipette
2 beakers
Rubber pipette attachment

Reagents

Ringer's solution
0·002 M KCN

Biological

Pithed frog

Determination of the mean osmotic concentration of plant cells at incipient plasmolysis

Osmosis is the chief means by which water is taken into a plant. In a plant cell there are two factors which decide whether water is taken in or given out. There is the dissolved material in the vacuole which determines the osmotic pressure of the cell sap and there is also the elastic cell wall which exerts a wall pressure on the cell contents.

Measurements of the osmotic pressure are usually made at incipient plasmolysis, that is when plasmolysis is just visible in about half the cells being examined.

Procedure

1. Mount thin sections of the tissue in sucrose solutions of various concentrations. Epidermal strips of leaves are suitable.
2. Leave for 20 minutes.
3. At the end of this time mount each piece of tissue on a slide in the appropriate sucrose solution and cover with a coverslip.
4. Observe about 25 cells in each tissue. Decide which solution is causing incipient plasmolysis.

Question

What can you conclude about the relative strengths of the internal and external solutions when about half the cells are just plasmolysed?

Requirements

Apparatus

Watch-glasses
Microscope slides and coverslips
Microscopes

Reagents

Series of sucrose solutions of molarities ranging from about 0·9 to 0·1

155

Biological

Tradescantia leaves
Elodea and moss leaves are suitable
because they are only one cell thick

Determination of the mean suction (water) pressure of plant material by the increase in length method

The uptake of water depends on the difference between the osmotic pressure and the wall pressure. This is the suction or water pressure.

Procedure
1. Cut several long thin strips out of a potato with the finest cork borer obtainable. The strips should all be made the same known length.
   They must all be cut along the same axis of the potato.
2. Place one in each of a series of sucrose solutions as in the previous experiment. Petri dishes are suitable for this.
3. After 1 hour measure the length of each strip.

Results
Plot a graph of the initial length: final length ratio against the molarity.

Question
What is the water pressure of the tissue?

Alternative experiment
Determine the weight of the tissues instead of the length.
Use the largest cork borer obtainable to cut large discs of tissue.
Adjust the size of these discs so that they each weigh about 2 g.
Dry them between filter paper and place in the same series of solutions as above.
Calculate the water pressure as above.

Requirements
Apparatus

Petri dishes with lids
Cork borer of two extreme diameters
Ruler and balance

Reagents

Series of sucrose solutions as in the previous experiment

Biological

Potatoes

156

The effect of temperature on the uptake of water by plants

Procedure

1. Cut 8 long thin strips of potato tissue with the finest diameter cork borer obtainable. Trim each strip so that they are all the same length. Measure the length.
2. Bring 8 boiling tubes of de-ionised water to the following temperatures by immersing them in water baths: 0°C or the temperature of an ice bath, and then the others should be 10, 20, 30, 40, 50, 60, 70 and 80 degC higher.
3. Place one strip in each tube.
4. Measure the length of each strip at 20 minute intervals.

Results

Plot a graph of strip length against time for each of the temperatures.

Determine the Q_{10} value between some of these temperatures from a graph of the rate of change of length of the strips at zero time against temperature. To do this you will first have to plot a graph of the rate of change of length against time and extrapolate each of these graphs back to zero time.

Measure the rate in change of length during each 20 minute period as a percentage of the original length.

Questions

1. What conclusions would you draw from the shape of your first graph?
2. What conclusions do you draw from the Q_{10} values you obtain?

Requirements

Apparatus

8 boiling tubes, each containing sufficient de-ionised water to cover the potato strips. Each of these tubes should be placed in a water bath at a different temperature, namely, iced water, and 10, 20, 30, 40, 50, 60, 70 deg higher
Ruler
Cork borer of very fine diameter

Biological

Potatoes. Cut all the strips along the same axis of the potato

The uptake of ions of opposite charge by plant material

Procedure

1. Cut 100 carrot discs, remove the excess water with blotting paper, and weigh them.
2. Pour 200 ml of 0·01 N ammonium chloride in a Dreschel bottle or a conical flask fitted with inlet and outlet tubes.
3. Attach the bottle or flask to a filter pump and draw a stream of air through the apparatus throughout the experiment. Check that the inlet tube actually passes almost to the bottom of the bottle. In this way all the carrot discs will be well aerated.
4. After 1 hour, 2 hours, 24 hours and 25 hours, remove two 5 ml samples with a 5 ml pipette. Use one of these samples to determine the concentration of ammonium ion and the other to determine the concentration of chloride ion.
5. **Determination of chloride ion concentration.**
   Add a few drops of potassium chromate to the 5 ml sample to act as indicator.
   Titrate silver nitrate solution into the sample. The end point is a faint permanent pink colour (after shaking).
   Repeat the titration with 5 ml of the original ammonium chloride solution.
6. **Determination of ammonium ion concentration.**
   Dilute the sample to 45 ml with de-ionised water.
   Add 2·5 ml of Nessler's reagent and make the volume up to 50 ml with more de-ionised water.
   Repeat this with 5 ml of the original solution.
   Place a portion of each of these solutions in a 3″ × 1″ specimen tube.
   Stand these tubes on a white tile alongside each other.
   Adjust the levels in the tubes until they both have the same intensity of colour when viewed from above.
   Measure the height of solution in each tube. The original solution would contain more ammonium ion if both levels were the same and therefore it would have an intense colour.

Results

To find the amount of ammonium ion taken up

The original solution will contain $0·18 \times 200/1{,}000$ g of NH_4^+.
After 1 hour the amount remaining in the solution in the bottle will be $0·18 \times 200/1{,}000 \times a/b$ where a represents the height of liquid in the tube from the 1 hour sample and b the height from the original solution.
Therefore the amount taken up in the first period
$$= (0·18 \times 200/1{,}000) - (0·18 \times 200/1{,}000 \times a/b) \text{ g}$$
Express the result in mg.

158

In calculating the amount taken up in subsequent periods, remember that you are starting with less than 200 ml of solution, e.g. after 24 hours, you only have 180 ml in the bottle (two 5 ml samples have been taken previously, one after 1 hour and the second after 2 hours).

Therefore the amount taken up between the second hour and the twenty-fourth

$$= (0.18 \times 190/1{,}000 \times c/d) - (0.18 \times 180/1{,}000 \times e/f) \text{ g}$$

where c, d, e and f represent the heights of solutions treated with Nessler's reagent.

To find the amount of chloride ion taken up

A normal solution contains 35·5 g of chloride per litre.
Therefore a 0·01 N solution contains 0·355 g per litre.
Therefore 200 ml contains 0·071 g.
After 1 hour the amount of chloride remaining $= (a/b \times 0.071) \text{ g}$ where a and b represent the titration reading with the 1 hour sample and the original solution sample.
Continue the calculation as for the ammonium ion.
Plot a graph of the amount of the ions taken up against time.
Express the ions as mg ions per g of carrot tissue.

Questions

1. What would you conclude from the shape of the graphs?
2. Can you offer any explanations for the difference in uptake of these two ions?

Requirements

Apparatus

Dreschel bottle or conical flask
fitted with inlet and outlet tubes
The inlet tube must reach almost to
the bottom of the flask
Balance
Filter pump
Burette
5 ml pipette
Conical flask for the titration
2 3 in × 1 in specimen tubes
White tile

Reagents

0·01 N NH_4Cl
Nessler's reagent
Silver nitrate solution
Potassium chromate

Biological

Carrots

159

The use of radioisotopes in experiments

Units

Curie

This is the number of disintegrations occurring in 1 g of pure radium per second, or as sometimes expressed, $3 \cdot 7 \times 10^{10}$ disintegrations per second (dps). This figure is arbitrarily chosen.

Millicurie (mc) is $1/1,000$ c or $3 \cdot 7 \times 10^7$ dps.

Microcurie (μc) is $1/1,000,000$ c or $3 \cdot 7 \times 10^4$ dps.

Note that the curie refers to the number of disintegrations actually taking place, rather than the number actually detected which is usually only a fraction of those taking place.

Precautions to be taken when handling radioisotopes

1. Do not allow materials to come in contact with your body. Do not eat in the laboratory.
   Wear an overall and rubber gloves. After each experiment wash your hands thoroughly and place them near the Geiger counter to see if any radioactive contamination is present. If necessary wash your hands again thoroughly.
2. **Never use a mouth pipette.**
3. To prevent contamination of the bench cover it with a sheet of polythene.
   If any radioactive material is spilt on the bench itself, wipe it at once with a soap solution and paper towelling. Dispose of this in a special waste bin. Flush the contents down the toilet. The polythene sheet must be thoroughly washed at the end of the experiment.
4. At the end of the experiment all the equipment used must be thoroughly washed and then examined with the Geiger counter and rewashed if necessary.
5. Place all material to be disposed of in the special waste bin.
6. It is wise to keep the apparatus used for these experiments just for these uses and to label it accordingly.
7. Particularly difficult stains and deposits of radioactive material may be removed with nitric acid.

Autoradiographic technique

This is used to investigate the sites of absorption of the radioactive material. The material containing the radioactive substance is placed on the photographic film and then after a period of time the film is developed. The areas which have become darkened represents the pattern of the radioactive substance.

This technique can be used to determine the localization of the

160

radioactive substance actually in the biological material or in chromatograms prepared from the material.

To autoradiograph the plant material, place it directly on top of the sensitive surface of a sheet of the X-ray film (suitable films are Kodak Industrex type D, Kodak Kodirex and Kodak Crystallex). This should be done in the dark room.

Seal between two sheets of cardboard with rubber bands.

Place on the bench with heavy weights on top to ensure contact between the material and the film at all points.

If the plant material is full of sap, place it on a piece of paper to which it should be fastened with Sellotape, and place the paper in a polythene bag. This will prevent the radioactive juices contaminating the film. Then place the bag on the sensitive side of the film with the plant material adjacent to the film.

Leave the plant material in this position for between 36 and 48 hours.

After exposure, develop the film in Kodak D–19b for 5–12 minutes at 20°C.

Fix in Kodak Unifix.

To autoradiograph a chromatogram, after spraying the chromatogram with locating reagent, and marking and identifying the spots, cover the chromatogram with polythene and set it up on the film in the dark room as for the plant material.

Leave it for 14 days before developing the film.

The Geiger counter

This can be used to detect radioactive substances. It can be passed over the surface of the growing plant and hence in this way assimilation of isotopes can be detected as it actually happens.

The actual amount of isotope can be determined and the amounts in different parts of the plant compared.

Similarly a chromatogram can be investigated by marking it off in $\frac{1}{2}$ cm sections and then determining the activity of each section with a Geiger counter with a rectangular limiting slit made from cardboard (0.5×2.5 cm) placed in front of the counter window. The activity of each section is then plotted against its position on the chromatogram.

A suitable Geiger counter is the Mullard MX 168 or the MX 168/01.

NOTE. Permission to use radioactive materials in schools can only be granted by the Department of Education and Science, Curzon Street, London, W.1. This is usually only granted when the teacher has attended an approved course of instruction.

The use of a radioactive tracer to study the uptake of a mineral element

Procedure

1. Remove a bean plant from the sand in which it is growing and wash the roots under the tap.
2. Place it in a flask containing labelled disodium hydrogen orthophosphate.
3. Cover the flask with lead foil to prevent the Geiger counter detecting this P^{32}.
4. Measure the activity in each part of the plant every 15 minutes.
5. After 2 hours remove the plant from the flask, cut off the roots and autoradiograph the plant.

Requirements

$Na_2HP^{32}O_4$. Supplied by J. A. Radley of Reading (220 Elgar Road)

Conical flask large enough to take the bean plant

Lead foil

Geiger-Muller tube

Bean plants growing in sand

The rate of output of contractile vacuoles in relation to the degree of osmotic stress

It is thought that the contractile vacuoles of many Protozoa control the body volume and the solute concentration of the body. They do this by bailing out the water as fast as it enters the body by osmosis. Otherwise the body would swell and the internal concentration would decrease.

If this view is true, the external osmotic pressure may be expected to bear some relationship to the activity of the vacuole. For experimental work it is most convenient to use sessile organisms.

Procedure

1. Place the animal on a slide under a coverslip which is supported at each end by a piece of filter paper.
2. Record the time at which the vacuole contracts in the normal culture for a number of successive contractions.
3. Estimate the diameter of the vacuole immediately before a contraction. Use a graticule for this purpose. This must first be calibrated (see the section on microscopy, page 9).

4. Irrigate the animal with 0·1 M ethylene glycol.
   Wait until the animal is again functioning normally, and then again measure the rate of contraction and the size of the vacuoles.
   Look for changes in the body size.
5. Irrigate with water and repeat the observations on the rate of vacuole contractions, size of vacuole, and body size.
6. Using a normal animal again, irrigate with a succession of dilutions of sea water, 2%, 4%, 6%, 8%, . . . , etc. until vacuolar activity ceases.
   Make observations as above.
7. Try the effect of non-electrolytes, for example sucrose solutions ranging about 0·05 M concentrations.

Results

Calculate the mean vacuolar volume when it is about to collapse in each solution. $v = (\pi \times \text{diameter}^3)$ cubic micra.
Calculate the rate of vacuolar output in μ^3/sec.
Plot a graph of rate of output against the concentration of the external medium.
Plot a graph of the volume of the animal against the concentration of the external medium.

Requirements

Apparatus

Microscope and bench lamp. Place a glass tank of copper sulphate between these to prevent overheating the animal
Coverslips and filter paper
Graticule and calibrated slide

Reagents

0·1 M ethylene glycol
Sucrose solutions around the 0·05 M range
Sea water of the following strengths: 2, 4, 6, 8% . . . etc.

Biological

Vorticella or *Carchesium*

Protoplasmic streaming

Procedure

1. Mount *Paramecium* in 10% methyl cellulose solution.
   Cover with a coverslip and observe under high power.
   Look for streaming in the region of the gullet.
2. Remove a leaf from near the tip of an *Elodea* plant and place on a slide in a drop of water.
   Cover and examine under high power.

3. Remove a piece of the epidermis of a nettle leaf and mount in water on a slide so a hair can be clearly seen.
   Cover and examine under high power.

Draw a cell of each specimen, indicating the direction of movement of cytoplasm.

Question
What is the purpose of protoplasmic streaming?

Requirements

Apparatus

Microscope: phase contrast if available
Microscope slides and cover glasses

Reagents

10% methyl cellulose

Biological

Paramecium
Elodea
Nettle

A study of the digestive and feeding mechanisms in a variety of animals

The purpose of this section is to show that the diverse feeding habits of many living forms leads to variations in digestive structure and physiology.

'Paramecium'—a demonstration of intracellular digestion

The path of the food
Prepare a wet mount of *Paramecia* in a drop of 10% methyl cellulose.
This tends to slow down the movement of the *Paramecia* because it has a higher density than water. The addition of some decaying material from the culture will help to support the coverslip and secondly the *Paramecia* will tend to gather around it.

Add a little powdered carmine to the drop and place a coverslip on it. Trace the path of the carmine grains into the oral groove and through the gullet into a food vacuole.

The food vacuoles can also readily be seen if *Paramecia* are irrigated on a slide with Indian ink or very dilute neutral red.

Digestion

Stain a suspension of yeast with Congo red. Place 1 drop of the yeast suspension on a slide with a drop of *Paramecia* culture. Observe the ingestion of yeast cells and the formation of food vacuoles.

Defaecation

This is best seen if the preparation is flooded with Indian ink.

Questions

1. How did the particles get into the animal?
2. How was the food vacuole formed?
3. What was the path of the food vacuole?
4. What was the colour of the yeast cells at the time of ingestion?
5. What was the colour of the yeast cells during their transport through the animal?
6. How does the passage of cytoplasmic materials into the vacuole affect the pH of the vacuole?
7. What evidence do you have that digestion has taken place?

In answering these questions note that Congo red is blue-violet at pH 3 or less, changing to purple above 3 and to red above pH 5·1.

'Hydra'—nematocysts

Excitation

1. Place *Hydra* in a cavity slide in pond water.
   Touch a tentacle with a clean coverslip; stain the coverslip with 1% methylene blue, and examine under the microscope.
2. Coat a coverslip with saliva, dry and repeat the experiment.
3. Place *Hydra* on a slide under a supported coverslip; irrigate with either 5% sodium chloride solution or 1% acetic acid.
   The examination of the discharged nematocysts may be facilitated by staining with 1% methylene blue.
4. Place crushed *Daphnia* near *Hydra* on a slide and observe the responses with a microscope. Remove *Daphnia* before it is eaten and stain for examination. What sort of nematocysts are attached to it?
5. Make up a 10% solution of saliva and place *Hydra* in it.
   Touch a tentacle with a fine glass rod (0·2 mm diameter).
   Compare the effect with the response obtained from *Hydra* which had not previously been treated with saliva.

Staining

1. Protruded nematocysts: acetic-methylene blue.
2. In a whole mount: kill with mercuric chloride (saturated aq.) at 60°C.
   Wash thoroughly in alcohol.
   Stain with methylene blue.
   Dehydrate rapidly.
   Clear in cedar wood oil.
   The nematocysts will be stained blue and the rest of the body will be unstained.

'Daphnia'

Stain a suspension of yeast cells with neutral red.
Add a drop of this to a cavity slide containing *Daphnia*.
Do not disturb for about 5 minutes.
Examine under the low power of the microscope. Focus on the gut along the dorsal length of the animal.

Questions

1. What evidence is there that digestion is taking place?
2. Where does it take place?
3. Is digestion extracellular or intracellular?
4. Why is extracellular digestion a better adaptation for *Daphnia*?

In answering these questions, note that the neutral red is red at pH 6·8, light red at pH 7·2 and yellow at pH 8·0.

Cockroach and locust

Mouth part preparation

1. **For a permanent preparation**, boil the whole head in 10% sodium hydroxide for 5 minutes. **Take care.**
   Wash in water.
   Dehydrate in alcohol (30%, 50%, 70%, 90%, 100%) 2 minutes in each, or in cellosolve.
   Clear in xylene.

Mount in micrex; either mounting the whole head or removing the mouth parts and mounting them separately.

2. **For a quick temporary preparation,** boil in 2% sodium hydroxide to remove the flesh from the mouth parts. This should take no more than 5 minutes. Then mount.

Food preferences

Mark a piece of filter paper 0·008 M, another 0·15 M, and another 0·30 M.

Dip these pieces of paper in sucrose solutions of the appropriate strength.

As a control soak some paper in de-ionised water.

Dry the papers in an oven.

Starve 16 locusts overnight.

Then feed 4 on each type of paper.

After 24 hours collect, count, and dry and weigh the faecal pellets. The weight of faeces produced indicates the amount of food eaten.

Palpal chemical sense

Observe the mouth parts of a hungry locust while holding it under the low power of a binocular microscope.

Prepare pieces of filter paper as in the above experiment and hold a strip at a time next to the labial and maxillary palps.

Observe the movements of the mouthparts and the fore-limbs.

Use a different locust for each paper.

Digestion

Dissect out the gut including the salivary glands.

Separate the salivary glands, crop, gizzard, mid-gut and hind-gut, transferring each to a separate solid watch-glass.

Cut each organ open and wash out the contents with water.

Grind each organ up in 2 ml of water.

Use these extracts to test for the presence of enzymes as follows:

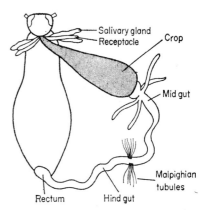

(a) **Invertase.** Place 3 drops of each extract in a separate labelled tube.

Add 1 ml of 5% sucrose to each.

Set up a control with water in place of the extract.

Incubate at 35°C for 1 hour.

Set up another set of tubes with 4 drops of water and 2 drops of Benedict's reagent.

Add a drop from one of the experimental tubes to one of the tubes of Benedict's reagent.

Heat over a water bath.

Repeat with each experimental tube and the control tube, noting which extracts show the presence of invertase by producing a red precipitate in the previously blue reagent.

(b) **Amylase.** Place 5 drops of each extract in a separate labelled tube.

Add 1 ml of a 1% solution of soluble starch to each tube.

167

Set up a control with water in place of the extract.

Incubate at 25°C.

At 5 minute intervals place a drop from each tube on to a spotting tile.

Add a drop of iodine solution to each spot, noting the extracts which show evidence of amylase activity.

(c) **Protease.** To demonstrate this it is necessary to pool extracts from ten insects.

Grind up the salivary glands of 10 insects together.

Repeat for each organ in turn.

Add 2 drops of each extract to labelled tubes.

Stain fibrin with Congo red.

Sprinkle a small amount of this stained fibrin into each tube.

Set up a control using water in place of extract.

Leave overnight.

Observe colour of stain.

Examine samples under microscope.

Cockroaches and locusts are herbivores; this experiment should also be performed with a carnivorous insect such as beetle or dragonfly larvae.

(d) **Lipase.** Each of the organs from 5 insects should be pooled in 5 ml of water.

Grind up as above.

It may be necessary to neutralise the extracts with acid or alkali.

Add 1 ml of olive oil to each extract.

Add 4 drops of sodium taurocholate (bile salts) to each extract.

Remove 0·5 ml from each extract and titrate it against 0·005 M sodium hydroxide, recording the volume required to convert the phenolphthalein indicator pink.

Incubate the remaining samples at 25°C.

At 30 minute intervals remove 0·5 ml samples for titration.

Continue for 3–4 hours if possible.

Goldfish—a herbivore

Dissect a freshly killed fish under water.

To do this cut through the ventral body wall with a pair of scissors.

Expose the coiled intestine and lift it out of the abdominal cavity.

Keep constantly moist with water.

Remove fat, mesentery, blood vessels.

Observe the extreme length of the intestine and compare its length with that of the rest of the alimentary canal.

Enzyme extract preparation

Cut open the intestine and wash out its contents with de-ionised water.

Place the intestine in a mortar and grind in 20 ml of 50% glycerol. Decant the liquid extract into a bottle, add another 20 ml of 50% glycerol to the residue.

Continue grinding this residue until you obtain a fine suspension. Pour this into a glass-stoppered bottle and add several ml of toluene, as a preservative, to each bottle.

Leave the bottles away from direct sunlight and at room temperature for 4 days.

Use these enzyme extracts to test for the presence of sucrase, amylase, and protease as described elsewhere.
Each time use a boiled control.

Mammals—digestion

Preparation of enzyme extracts

1. **Salivary gland and pancreas.** Rinse the organ under the tap.
Dry on a piece of filter paper.
Grind with sand in a mortar.
Add 5 ml of de-ionised water and continue grinding.
Pour the mixture into a test tube and shake vigorously for 10 seconds.
Before using any extract, mix well by inverting.
Or, rinse a mouthful of tap water well around the mouth and spit it into a test tube.
Shake it up and filter through a small wad of cotton wool.

2. **Stomach and small intestine.** Cut the organ open along its whole length.
Wash it out under the tap and dry on filter paper.
Scrape off the mucous membrane with a scalpel.
Transfer the scrapings to a tube.
Add the required amount of acid or water.
Shake vigorously for 10 seconds.
Before removing any extract, invert to mix.

Salivary amylase

1. Set up test tubes as follows:
   Tube 1. 1 ml 1% starch, 1 ml saliva. Set this tube up 3 times
   Tube 2. 1 ml 1% starch, 1 ml saliva, 1 drop concentrated HCl
   Tube 3. 1 ml 1% starch, 1 ml saliva (boiled)
   Boil the enzyme by heating a small amount in a beaker of boiling water for 3 minutes.

2. Mix each tube well and incubate at 37°C for 10 minutes.
3. Prepare 3 test tubes, each containing 3 ml of de-ionised water and 1 drop of iodine solution.
4. After incubating, transfer 1 drop in a pipette from tube 1 to one of the iodine tubes. Mix. (only use one tube 1).
5. After rinsing the pipette, repeat the test with tubes 2 and 3.
6. Using samples which have given a negative test for starch, test for the presence of maltose and glucose.

Maltose test—to 3 ml of the test solution add 3 drops of 20% NaOH and 3 drops of 5% methylamine hydrochloride.
Shake. Boil in a beaker of boiling water for 30 seconds.
Remove at once and shake.
With maltose, a deep carmine colour appears.
With glucose, a deep yellow colour appears. This is the response of a reducing sugar. Confirm by adding 2 drops of Benedict's reagent to the hot mixture. A dense orange precipitate appears.
Sucrose test—use the third tube 1. Add 1 drop of concentrated HCl, mix and heat to boiling. Make alkaline and test for glucose.

Influence of chloride ions on salivary amylase

1. Dilute 1 ml of undiluted saliva up to 10 ml with de-ionised water.
2. Shake and filter through a small wad of cotton wool.
3. Set up test tubes as follows:
   Tube 1. 5 ml of 1% starch solution, 1 ml of 0·9% NaCl.
   Tube 2. 5 ml of 1% starch solution, 1 ml of de-ionised water.
4. Place both tubes in a water bath at 37°C.
5. Place several spots of iodine solution on a spotting tile.
6. Add 1 ml of the filtered, dilute saliva to tube 1 and another 1 ml to tube 2.
7. Mix each tube and immediately return to the bath, noting the time.
8. At intervals of 1 minute transfer a drop from tube 1 to an iodine spot. Mix.
   The colour is first blue, and then violet, and finally no colour is obtained.
   Note the time taken for this change.
9. Then repeat the procedure at 2 minute intervals with drops from tube 2.
10. Compare your results with those of other people.

Pepsin

1. Prepare a suspension of the mucous membrane of the stomach in 5 ml of 0·05 N—HCl (pH 1·3).
2. Transfer 2 ml to a test tube.
3. Heat a second 2 ml sample in a beaker of boiling water for 3 minutes.
4. Cool under the tap.

5. Add a small disc of egg white to each tube.
6. Incubate at 37°C overnight.
7. Observe the degree of turbidity.

Activation of trypsinogen by enteropeptidase

1. Prepare suspensions of the pancreas and the mucous membrane of the proximal 2 cm of the small intestine, both in 5 ml of de-ionised water.
2. Set up test tubes as follows:
    Tube 1. 2 ml barbitone-HCl buffer (0·1 M, pH 8·0).
            1 ml pancreatic suspension (trypsinogen).
    Tube 2. 2 ml barbitone-HCl buffer.
            1 ml intestinal suspension (enteropeptidase).
    Tube 3. 1 ml barbitone-HCl buffer.
            1 ml pancreatic suspension.
            1 ml intestinal suspension.
3. Add a small spoonful of Congo red fibrin powder to each tube.
4. Shake.
5. Incubate at 37°C for 20–30 minutes.
6. Shake the tubes every 5 minutes.
    Tubes 1 and 2 should be negative (colourless) or only slightly positive (faintly pink). Tube 3 should be positive (deep pink to red).

Activation of chymotrypsinogen by trypsin

1. Set up test tubes as follows:
    Tube 1. 1 ml de-ionised water, 1 ml intestinal suspension (enteropeptidase).
    Tube 2. 1 ml de-ionised water, 1 ml pancreatic suspension (trypsinogen and chymotrypsinogen).
    Tube 3. 1 ml pancreatic suspension, 1 ml intestinal suspension.
2. Mix and incubate at 37°C for 10 minutes.
3. Add 1 ml of 0·2 M acetate buffer (pH 5·0).
4. Add 2 ml of milk.
5. Mix and incubate at 37°C for 10 minutes.
6. Observe which tubes contain clotted milk.
    If trypsin and enteropeptidase do not clot milk, how is it clotted here?

Pancreatic amylase

1. Prepare a suspension of the pancreas in 5 ml of de-ionised water.
2. Heat half of this in a beaker of boiling water for 3 minutes.
3. Cool under the tap.
4. Set up test tubes as follows:
    Tubes 1 and 2. 1 ml phosphate buffer (0·1 M, pH 7·0).
                1 ml 1% starch solution.
                1 ml pancreatic suspension.
    Tubes 3 and 4, as for tubes 1 and 2 but using boiled suspension.

171

5. Mix and incubate at 37°C for 10 minutes.
6. Test tubes 1 and 3 for starch with iodine solution.
7. Test tubes 2 and 4 for reducing sugars with Benedict's reagent.

Pancreatic lipase

1. Prepare the suspension as for the amylase experiment.
2. Set up test tubes as follows:

   Tube 1. 1 ml milk, 3 drops of methyl red, 1 ml pancreatic suspension.
   Tube 2. 1 ml milk, 3 drops of methyl red, 1 ml boiled pancreatic suspension.
3. Mix and incubate at 37°C.
4. Observe change in colour in tubes, and rancid odour.
5. Add 0·1 N NaOH to the tube which has changed colour until its colour matches that of the control.
   The amount of NaOH added is a measure of the fatty acid produced.

Maltase

1. Prepare a suspension of the mucous membrane of the proximal 10 cm of the small intestine in 5 ml of de-ionised water.
2. Heat half of this in a beaker of boiling water for 3 minutes.
3. Set up test tubes as follows:
   Tube 1. 1 ml phosphate buffer (0·1 M pH 7·0).
           1 ml intestinal suspension, 1 ml of 1% maltose.
   Tube 2. As for tube 1 but using boiled suspension.
4. Mix and incubate at 37°C for 5 minutes.
5. Test for glucose. This must be a specific test as both glucose and maltose are reducing sugars. Use a Clinistix strip. Dip this in both solutions for about 5 seconds.
   If glucose is present a blue colour develops.

Sucrase

1. Prepare a suspension of the mucous membrane of the second 10 cm in 3 ml of de-ionised water.
2. Heat 1 ml in a beaker of boiling water for 3 minutes.
3. Cool under the tap.
4. Set up test tubes as follows:

   Tube 1. 1 ml of phosphate buffer (0·1 M, pH 7·0).
           1 ml 1% sucrose.
           1 ml intestinal suspension.

   Tube 2. As for tube 1 but using boiled suspension.
   Mix and incubate at 37°C for 15 minutes.
   Add 0·5 ml of Benedict's reagent to each tube.
   Mix and heat in a beaker of boiling water for 2 minutes.
   Observe where the reagent is reduced.

172

Requirements
Apparatus

Cavity slides
A microprojector and a phase contrast microscope will be found useful for showing *Paramecia*, nematocysts and *Daphnia*
Low-power phase is particularly useful for observing nematocysts discharging
Solid watch-glasses
Water baths
Mortar and pestle with silver sand
Spotting tiles

Reagents

For use with *Paramecium*—10% methyl cellulose
Powdered carmine
Congo red
For use with *Hydra*—1% methylene blue
5% NaCl
1% acetic acid
Acetic-methylene blue
Mercuric chloride (sat. aq.) alcohol
Cedar wool oil
For use with *Daphnia*—neutral red
For use with insects—10% NaOH, 2% NaOH
Alcohol, 30, 50, 70, 90, 100%
Cellosolve
Xylene
Micrex
Sucrose, 0·008 M, 0·15 M, 0·30 M
Benedict's reagent
1% soluble starch
Congo red

Fibrin
Olive oil
sodium taurocholate (bile salts)
For use with goldfish—50% glycerol
Toluene
For use with mammal—1% starch
Conc. HCl, 0·05 N HCl
20% NaOH, 0·005 M NaOH
5% methylamine hydrochloride
Benedict's reagent
0·9% NaCl
Barbitone-HCl buffer (pH 8·0)
Congo red
Fibrin
Acetate buffer (pH 5·0)
Milk
Phosphate buffer (pH 7·0)
iodine/KI solution
Methyl red
1% maltose solution
1% sucrose solution
Clinistix strips

Biological

Paramecium
Yeast
Hydra
Daphnia

Cockroaches or locusts
A small mammal
Goldfish

The investigation of factors affecting photosynthesis

The purpose of this experiment is to study the effect of varying the external environment of a plant in an attempt to discover the optimum conditions for photosynthesis.

Procedure

Either use a simple manometer as shown in diagram or an Audus microburette as shown in diagram. **When using the manometer** with

173

Elodea, pour sodium bicarbonate solution into both test tubes until they are ¾ full. Then add a sprig of *Elodea* to one of the tubes. If using algae, centrifuge the cells out of the culture solution and pour off the solution. Add sodium bicarbonate to the cells. ¾ fill one of the tubes with this mixture, adding more bicarbonate if necessary. Pour an equal amount of bicarbonate in to the other tube.

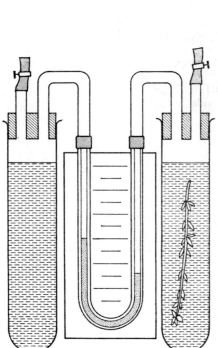

SIMPLE
MANOMETER

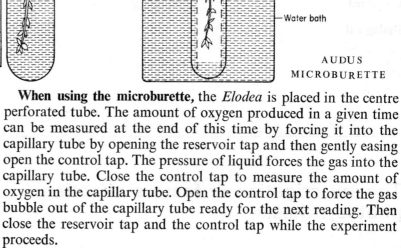

AUDUS
MICROBURETTE

When using the microburette, the *Elodea* is placed in the centre perforated tube. The amount of oxygen produced in a given time can be measured at the end of this time by forcing it into the capillary tube by opening the reservoir tap and then gently easing open the control tap. The pressure of liquid forces the gas into the capillary tube. Close the control tap to measure the amount of oxygen in the capillary tube. Open the control tap to force the gas bubble out of the capillary tube ready for the next reading. Then close the reservoir tap and the control tap while the experiment proceeds.

In both pieces of apparatus the outer beaker is used for temperature regulation by adding to it small amounts of hot or cold water.

Determination of the effect of light intensity on the rate of photosynthesis

Allow 5–10 minutes for the tubes containing the plant material to come to the temperature of the water bath.

174

For both methods the outer beaker should be painted black with one vertical, narrow slit unpainted through which light can pass. This ensures that the plant is only receiving light from the known source of light.

Measurements should be taken using the following light intensities:

(a) No light. Seal out the light by wrapping the tubes in aluminium foil.

(b) 32 cm from the test tube containing the plant.

(c) 20 cm from the test tube containing the plant.

(d) 14 cm from the test tube containing the plant.

(e) 10 cm from the test tube containing the plant.

At each light intensity, several readings should be taken at 2 minute intervals to give an average for the oxygen production, (microburette experiment) or average rate of pressure change (manometer experiment).

Determination of the effect of temperature on the rate of photosynthesis

The apparatus should be used as in the previous experiment but the intensity of light should be kept constant.

The experiment should be performed twice, once with a low light intensity and once with a high light intensity.

In each experiment the rate of photosynthesis should be measured at 20°C and 30°C or two other temperatures 10°C apart.

Determination of the effect of various concentrations of carbon dioxide

The apparatus should be used as in the previous experiments.

The intensity of light and the temperature should be kept constant.

Measurements should be taken using the following concentrations of sodium bicarbonate: 0·01, 0·05, 0·10, 0·15, 0·20, 0·30, 0·40. All these figures are expressed as % of $NaHCO_3$.

Determination of the effect of different colours of light on the rate of photosynthesis

The apparatus should be used as in the previous experiments. An unpainted beaker should be used as the water bath and the experiment should be performed in the dark room. Coloured lamps should be used or alternatively ordinary bench lamps covered with coloured Cellophane paper.

If a dark room is not available, the black-painted beaker should again be used and coloured solutions should be used.

In this case it is important that the light intensity is the same for each solution. This can be done by measuring the intensity of light transmitted by one coloured solution and then adding drops of another coloured dye to a second beaker of water until the meter registers the same reading.

175

Use red, blue, green and a control of white light.

As in the previous experiments all the other factors must be constant throughout the experiment and they must as far as possible be optimum.

Results

(i) Light intensity at distance d from the lamp is proportional to $1/d^2$, therefore a graph of $1/d^2$ against the amount of oxygen produced or average rate of pressure change (mm/min) can be plotted.

(ii) In general increasing the temperature of a chemical reaction produces an increased reaction velocity, and for a temperature rise of 10 degC, the velocity is doubled. This is called the temperature coefficient Q_{10} which is defined as

$$\frac{\text{The rate of a reaction at a temperature of } (t + 10)°C}{\text{The rate of reaction at a temperature of } t°C}$$

Calculate the Q_{10} for the high intensity light experiment and also for the low intensity light experiment.

(iii) Plot a graph of the concentration of carbon dioxide against the rate of oxygen production.

(iv) Plot an action spectrum, i.e. rate of photosynthesis against the approximate wavelengths of the lights used.

This should be compared with absorption spectra of the photosynthetic pigment and of its separate parts (page 183).

Questions

1. When studying the rate of photosynthesis with various light intensities, what factors are controlling the rate of reaction?
2. In the first experiment what reason have you for believing that other factors are controlling the rate of reaction?
3. What is the purpose for calculating the Q_{10} values at both high and low light intensities?
4. Which wavelengths of light are the most active in causing photosynthesis?
5. In conclusion what do all these experiments suggest about the nature of the photosynthetic reaction?

Requirements

Apparatus

Simple manometer or microburette
Black-painted beakers with slit
Lamps, including either coloured ones or coloured Cellophane paper
Metre ruler
Light meter

Reagents

$NaHCO_3$ solutions of the following strengths, 0·01%, 0·05%, 0·10%, 0·15%, 0·20%, 0·30%, 0·40%

Biological

Elodea or suitable algal culture such as *Chlorella*

176

The mechanism of stomatal movement

This experiment seeks to discover the nature of the forces which open and close stomata.

Procedure
1. Fill two watch-glasses with de-ionised water.
2. Fill another two with 33% calcium chloride solution.
3. Fill another two with 0·001 N hydrochloric acid.
4. Fill another two with 0·001 N potassium hydroxide.
5. Place a *Tradescantia* leaf in each solution with the lower epidermis uppermost.
6. Place one of each of the solutions containing a *Tradescantia* leaf in the dark for 15 minutes.
7. Place the other solutions under a bench lamp for 15 minutes.
8. Strip off the lower epidermis from a ninth leaf and place it in alcohol.
9. Place this piece of epidermis on a slide, cover with a coverslip and observe 25 stomata, counting the number open and the number closed.
10. At the end of the 15 minutes, remove the lower epidermis from each leaf and plunge each into a separate watch-glass of alcohol.
11. Take care to label each glass.
12. Observe 25 stomata and count the number open and closed for each leaf.

Results
Arrange the leaves in order according to the number of stomata which are open.

Questions
1. What can you say about the closure and opening mechanism of stomata from your results with calcium chloride?
2. In the light of observations with calcium chloride, what changes in the cells may the acid and alkali be causing?
3. Can you suggest a possible explanation for these results?

Requirements

Apparatus
Microscopes and slides
8 watch-glasses

Reagents
Alcohol
33% calcium chloride
De-ionised water
0·001 N HCl
0·001 N KOH

Biological
Tradescantia leaves

A comparison of the compensation periods of sun and shade plants

This experiment shows that light is an important ecological factor. The compensation period is the time a plant takes to make up by photosynthesis the loss of carbohydrates that occurs due to respiration.

In this experiment a plant is allowed to respire until an indicator shows the acid colour. Then it is illuminated and the time required to convert the indicator colour back to the original is noted. This is the compensation period.

Procedure

1. Line the bottom of two large crystallising dishes with filter paper.
2. Make up a 0·2% solution of sodium bicarbonate containing a little 0·2% phenol red indicator. This should be shaken for about an hour to allow it to equilibrate its carbon dioxide content with that of the atmosphere. It will be a pale red colour.

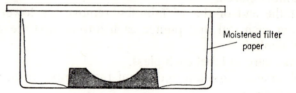

Moistened filter paper

3. Moisten the filter paper with de-ionised water.
4. Place 1 ml of the bicarbonate/indicator solution in a solid watch-glass and place it on the filter paper in the crystallising dish.
5. Arrange a shade plant around the watch-glass in one dish and a sun plant in the other dish. Only a minimum amount of soil should be used.
6. Seal each dish with a vaselined piece of glass.
7. Keep each dish in the dark for about 10 hours. The temperature of each dish must be the same and it must be kept constant.
8. Transfer both dishes to the light. Illuminate with bench lamps. The lamps must not heat the plant material and both plants must receive the same amount of light.
   At the end of the dark period the indicator will be yellow. The compensation period is the time taken for this colour to be restored to its original pale red.
9. Compare the colours by placing the dishes over a white tile.

Questions

1. What type of habitat will a plant which is slow in reaching its compensation point live in?

2. Under what conditions will a plant which is fast to reach its compensation point be at a disadvantage?
3. List the different types of plants which have short compensation periods.

Requirements

Apparatus

2 large crystallizing dishes
Filter paper
2 solid watch-glasses
Glass plate large enough to act as a lid to the crystallising dishes
Bench lamps

Reagents

0·2% NaHCO$_3$
0·2% phenol red
De-ionised water
Vaseline

Biological

Sun plant and a shade plant, i.e. *Taxaeum officinale* and *Mercurialis perennis*

Separation of photosynthetic pigments by paper chromatography

Procedure

Extraction of pigments

Mince the leaves in 90% acetone in a blender for two minutes, or grind them in acetone in a mortar.
Filter the extract through a Buchner funnel into a separating funnel.
Add an equal volume of petroleum ether (B.P. 100–120°).
Shake the mixture.
Wash three times in water, each time disgarding the lower aqueous layer.
Add solid sodium sulphate to the emulsion.
Leave the solution to stand over the sodium sulphate for some minutes.
While this is the ideal extraction method, good results can be obtained merely by grinding or blending the leaves in acetone, and then using the filtrate.

Solvents

100 parts of petroleum ether (B.P. 100–120°); 12 parts of 90% acetone.

Paper

25 cm × 25 cm paper is the most convenient as several spots can be made for each run.

Running time

About 6 hours should produce a good separation, although the paper could be left to run overnight. Run in diffuse light.

Developer

No developer is required, although as the pigments fluoresce in ultraviolet, they can be more easily seen in this light.

Results

Yellow band	0·95	carotene, contains *a* and *b*
Yellow-grey	0·83	phaeophytin, a breakdown product of chlorophyll which is not always clearly visible
Yellow-brown	0·71	xanthophyll, often appear as two bands
Blue-green	0·65	chlorophyll *a*
Green	0·45	chlorophyll *b*

Requirements

Apparatus

Blender or mortar and pestle
Separating funnel
Buchner funnel
Chromatography tank
Ultraviolet lamp

Reagents

90% acetone
Petroleum ether (B.P. 100–120°)
Solid sodium sulphate
25 cm × 25 cm filter paper

Biological

Nettles or ivy

Thin layer chromatography

Advantages of this technique

1. Filter paper has a rather coarse, fibrous nature and so the sample spots tend to spread out. On thin layers the spots remain quite compact and so much smaller amounts of substance can be separated.
2. This technique is much more rapid than paper chromatography.

Procedure

Preparation of plates

1. Microscope slides make ideal plates for some experiments, but any type of glass plate of similar thickness is suitable.
2. Arrange these slides on a frame so that they cannot move. It may be necessary to Sellotape their ends together. The slides must be absolutely clean. If it is necessary clean them in acetone.
3. Silica gel is used as the adsorbent. Make a slurry of the silica gel by adding 2 ml of de-ionised water for every 1 g silica gel. Alternatively chloroform can be used instead of water. The use of the latter is to be recommended as water reacts irreversibly with the gel and sets within 2–3 minutes. This inevitably leads to a wastage of adsorbent. When the plates have been prepared with water, they must be dried in the oven at 80–90°C for 30 minutes.

   If the slurry is made with chloroform, it will not set, and if evaporation losses are made good, it can be kept indefinitely. Having prepared the plates from this slurry, they are ready for use almost at once as the chloroform rapidly evaporates. 10 sq cm of plate requires 2·5 g of silica gel.
4. To pour the adsorbent over the plates a spreader is needed. This can take the form of a thick glass rod. If this is used stick Sellotape along the ends of the slides so that the spreader is raised slightly above the slides. Pour the adsorbent on to the slide at one end and roll it out along the other slides.

   A better way is to use a special applicator which is filled with the slurry and then drawn along the slides. The slurry is discharged from the bottom of the applicator and forms a film on the slide of a known thickness.

   An even more simple method, is merely to dip the slide into a bottle of the slurry. When it is withdrawn, it will be coated on both sides with a thin film of the adsorbent.
5. Even when chloroform is used the activity of the adsorbent can be enhanced by drying the slide in the oven for 2–3 minutes.

Solvents

These can be the same as for paper chromatography.

Application of sample to thin layer

A piece of very fine capillary tube must be used and the same technique should be adopted as for paper chromatography.

As the thin layer cannot be marked in the way that paper can, a mark must be scratched on either side of the slide at the level of the origins.

The spots should be 10–15 mm apart.

Operation

Any jar with a screw cap is suitable for use as the chromatogram tank. When microscope slides are being used, 125 ml specimen bottles with screw caps are suitable.

With the spots at the lower edge, pick up the slide by the edges and lower it carefully into the tank containing sufficient solvent to come up the slide to a point just below the origins.

If only one side of the plate is coated with adsorbent, two plates can be run simultaneously.

Take care not to place the slide into the tank at an angle as this may seriously affect the result if two spots are being compared as separation may be complete after a few minutes.

Drying the plates

When you take the plate from the tank, blot the lower edge and mark the solvent front with a needle.

Then oven or air dry the plate.

Locating reagents

In this book this technique is only used to separate readily visible pigments, but if a locating reagent is required, it **must be sprayed on to the plate.** The actual reagent used will be the same as for paper chromatography.

Preserving the records

When stable colours are obtained, they can be preserved by spraying the plate with a clear plastic emulsion. Polyvinylpropionate (Neatan) is suitable for this purpose.

When the plastic sheet is dry, peel it off the glass. The adsorbent will be embedded in a clear, flexible plastic sheet.

Requirements

Apparatus

Frame for supporting plates while spreading the adsorbent
Sellotape
A spreader, this may be a jar of suitable size to hold the adsorbent while dipping the slide in it
   or, a thick glass rod
   or, a special applicator. A suitable one can be bought from Shandon Scientific Company
Glass plates. 2·5 cm × 7·5 cm microscope slides are suitable
Alternatively window glass can be cut if larger plates are required
Oven

Reagents

Silica gel (Kieselgel G Nach Stahl. can be obtained from Camlab (Glass) Ltd, Cambridge
Chloroform Neatan (Polyvinylpropionate from Camlab)

Separation of photosynthetic pigments by thin layer chromatography

Procedure

Extraction of pigments
As for paper chromatography.

Solvents
25 parts of acetone; 75 parts of petroleum ether (B.P. 40–60°).

Running time
25 minutes. Work in diffuse light and if possible in a refrigerator.

Developer
No developer is required, although as the pigments fluoresce in ultraviolet light, they can be more easily seen in this.

Results
As for paper chromatography.

Requirements

Apparatus
Blender or mortar and pestle
Separating funnel
Buchner funnel
Small screw-top jar large enough to hold a microscope slide
Ultraviolet lamp
Microscope slides

Reagents
90% acetone
Petroleum ether (B.P. 40–60°)
Petroleum ether (B.P. 100–120°)
Solid sodium sulphate
Silica gel

Biological
Nettle or ivy

The absorption spectra of the chloroplast pigments

The purpose of this experiment is to compare the absorption spectra of the total extract and of its separate parts with the action spectrum for photosynthesis (see 'The investigation of factors affecting photosynthesis' page 173).

183

Procedure

The use of the direct-vision, hand spectroscope

It consists of two tubes, one of which has an adjustable slit at one end. The other, sliding within the first, has a focusing eyepiece and the prism.

Look through the spectroscope and turn it so that the red end of the spectrum is to the left.

Close the slit by turning the milled collar and then open it so that the colours of the spectrum are just visible.

Now focus by sliding the inner part in the outer until the fine black, vertical Fraunhofer lines are sharply defined.

Note the D line in the orange and the E line in the green.

These lines are due to the absorption of light by different elements in the sun's atmosphere, e.g. D line is due to sodium.

Any horizontal which may be seen are due to specks of dust on the edge of the slit.

The chloroplast pigments have the property of absorbing rays in various regions of the coloured spectrum. If solutions of these pigments are placed between the source of light and the spectroscope, the spectrum will be interrupted by a number of black areas.

It is possible to identify each substance from its absorption spectrum.

Extraction of pigments

Mince nettle leaves in 90% acetone in a blender for two minutes or grind them up in acetone in a mortar.

Filter the extract over a Buchner funnel.

Place the filtrate in a separating funnel and add an equal volume of petroleum ether (B.P. 100–120°) and shake the mixture.

Wash in water three times, discarding the watery lower layer each time.

Add solid sodium sulphate to help break down the emulsion.

Leave the solution to stand for some minutes over solid sodium sulphate.

The individual pigments can then be separated either by column chromatography or by using a variety of solvents and a separating funnel.

1. *Column chromatography*

Fill a 30 cm long, 1 cm diameter glass tube with pure Whatman cellulose powder. Use a small piece of glass wool as a pad at the lower end.

Only a little cellulose should be added at a time, and a ram-rod should be used to firm it down between each addition.

Wash the column with a mixture of acetone and petroleum ether (12 : 100 parts respectively). Use a filter pump to help draw the mixture through the column.

Add 5 ml of the pigment extract to the top of the column, and allow it to penetrate down the column.

When the pigments are beginning to separate, place a separating funnel containing the solvent at the top of the column and adjust it to drip steadily.

The least adsorbed pigments are first to appear at the bottom of the column. After a few minutes the carotene solution will begin to drip out. Collect it in a test tube.

The other pigments dripping out of the column in order will be phaeophytin, xanthophyll, chlorophyll *a* and *b*. Collect each in a test tube held under the column.

2. *Using solvents and a separating funnel*

5 ml of the petroleum ether extract should be shaken with 92% methyl alcohol in a small separating funnel.

Two layers form. The upper layer of petroleum ether contains chlorophyll *a*. It will be blue-green colour.

The lower layer of methyl alcohol contains chlorophyll *b*.

The layers will be to some extent obscured as the upper also contains carotenes and the lower the xanthophylls.

To separate the carotenoids, 5 ml of the original petroleum ether extract should be added to diethyl ether.

Then add 2 ml of 30% potassium hydroxide in methyl alcohol.

A brown colour develops which soon changes to green.

Add 10 ml of water slowly and then add a little di-ethyl ether.

Two layers will form; the upper will contain the carotenoids; and the lower the green pigments.

Run off the lower layer, and then wash the remaining upper layer with water.

Again run off the lower layer.

Evaporate the carotenoid extract in a basin down to 1 ml.

Dilute with 10 ml petroleum ether and an equal volume of 90% methyl alcohol.

Gently rotate the separating funnel to mix the liquids.

The lower layer of methyl alcohol contains xanthophyll.

This is a yellowish colour.

The upper layer of petroleum ether contains the carotenes.

This is a reddish-yellow colour.

Results

The best way to record your results is to mark vertical lines on a sheet of paper to represent the D and E lines. Then mark in the absorption bands in relation to these.

Question

Compare your absorption spectra with the action spectra obtained in the experiment to investigate the factors affecting photosynthesis. Which pigments absorb the most energy for photosynthesis?

185

Requirements

Apparatus

Direct vision hand spectroscope
Blender or mortar and pestle
Buchner funnel
Separating funnel, large and small
Ram-rod
Chromatography column (glass tube 30 cm long, 1 cm diameter)
Evaporating basin

Reagents

90% acetone, and G.P.R. acetone
Petroleum ether (100–120°)
Solid sodium sulphate
Pure Whatman cellulose or magnesium oxide powder
Glass wool
92% methyl alcohol
30% KOH
Diethyl ether

Biological

Any plant which is rich in chlorophyll
Nettles are ideal
Ground ivy
Frozen spinach

The formation of starch from sugars in the dark

This experiment shows that the final stage of photosynthesis, namely the production of the complex carbon compound, is independent of light.

Photosynthesis involves a dark stage as well as a light one. The light reaction generates reducing power by splitting the water molecule. The hydrogen ions so formed are passed to TPN via an unknown intermediate. This reaction is light-independent. This is then used to reduce the carbon dioxide fixation product, phosphoglyceric acid, ultimately to a three-carbon sugar.

Finally 6-carbon sugars are formed which are converted into higher carbohydrates. Once the reducing power has been generated these reactions are independent of light.

Procedure

1. Sterilise two beakers and glass plates which will act as lids.
2. Sterilise the experimental solutions, 5% sucrose and de-ionised water, and partly fill one beaker with sucrose and the other with water.
3. Remove two de-starched *Pelargonium* leaves from a plant.
4. Cut a new surface to the stalk close to the lamina.
5. Sterilise them by placing in 1% sodium hypochlorite for 3 minutes.
6. Wash them in sterile de-ionised water and float one on each experimental solution.
7. Cover each beaker with a piece of glass. This must permit plenty of air to enter the beaker so a tight joint is not necessary.

The covers are merely to prevent bacterial and fungal spores from entering the beakers.

8. Place the beakers in a dark cupboard for a few days.
9. At the finish, label the leaf which has been floating on sucrose by cutting a deep notch in it.
10. Place both leaves in a beaker of boiling water for a minute.
11. Then transfer to a beaker of 90% alcohol and **boil over a water bath.** This removes the chlorophyll.
    When the leaves are colourless they will be brittle and hard.
12. Place on a white saucer and moisten with tap water until they are limp again.
13. Pour off the water and immerse in iodine in potassium iodide solution.
14. When there appears to be no further colour change, wash off the excess stain.
15. Compare the colours of the leaves from the two solutions. Note any differences in the density of the colour in different parts of the leaf.

Requirements

Apparatus

2 sterilised beakers and glass plates which can act as lids
Third beaker
Boiling water bath
White saucer

Reagents

5% sucrose
1% sodium hypochlorite
90% alcohol
I_2/KI solution

Biological

2 de-starched *Pelargonium* plants

A study of the first products of photosynthesis

Although the order of formation of the earlier formed products of photosynthesis can only be shown with radioactive tracers, the later formed products can be demonstrated by convenient techniques. This experiment demonstrates the order of formation of glucose, sucrose and ribose.

Procedure

1. *Lemna* and *Elodea* should be kept in the dark for 24 hours before the experiment.
2. Controls which had been in the light all the time should also be used.

187

3. Working in the dark room, three 0·5 g samples of each of the plants were weighed out and each sample was placed in a separate Petri dish of water. This water should have been standing in the dark room for some time before the experiment to bring it to the temperature of the water the plants had previously been growing in. Only a faint blue light should be used while performing the weighing.
4. Each of the experimental samples should then be placed under a strong lamp.
5. After 5 minutes one sample of each should be plunged into boiling 90% alcohol. This kills the plant material. **This transfer must be performed very rapidly** as photosynthesis is continuing right up to the moment the plant enters the alcohol.

WARNING: The alcohol must be boiled over a water bath.

The sample can be kept in this state until required.

6. The other samples should be plunged into separate portions of boiling alcohol after 15 minutes and 30 minutes.
7. Each sample should be ground up in a mortar and filtered into an evaporating basin.
8. Evaporate the alcohol off **over a water bath.**
9. Add 1 ml of water to the residue and swirl it around so that it includes all the residue.
10. Spot this solution on to chromatogram paper.
11. Other spots should be made of ribose, sucrose, glucose, fructose and the control plant material.
12. Run the chromatogram in a phenol solvent. **Handle carefully.**
13. Develop the chromatogram in a 10% solution of resorcinol in acetone which contains a few drops of concentrated hydrochloric acid.
14. The spots will appear when the paper is dried.

Results

Compare the experimental results with those of the control and the known sugars.

Requirements

Apparatus

Balance
Dark room, with faint blue light
3 Petri dishes, each containing water which had been standing for some time
3 strong lamps
Beakers containing alcohol boiling over a water bath
Mortar and pestle
1 ml pipette
Capillary tubing
Chromatogram tanks and paper

Reagents

Phenol solvent 28 g of phenol and 12 ml of de-ionised water
10% resorcinol in acetone plus a few drops of HCl

Biological

Lemna or *Elodea*

A study of the fixation of carbon dioxide in the leaf using C^{14} radioactive tracer

See page 160 for radioactive tracer techniques and precautions.

Procedure

1. Place a pot plant on a ground glass plate under a belljar. Alternatively a sprig of a plant can be used in a small beaker of water as shown in the diagram. Set up as many sets of apparatus as possible.
2. Place sufficient labelled sodium carbonate in the apparatus to give a strength of 10 μc.
3. Place a 100 watt lamp about 40 cm from the plant material.
4. Generate the carbon dioxide from the sodium carbonate by the addition of acid from a dropping funnel. The addition of 2 M hydrochloric acid to 1 ml of labelled sodium carbonate will provide sufficient $C^{14}O_2$.
5. Illuminate the belljars for $\frac{1}{2}$ hour, 1 hour, $1\frac{1}{2}$ hours, 2 hours and 3 hours respectively.
6. Then open the belljar in a fume cupboard.
7. Autoradiograph the plant to discover the sites of absorption of the carbon dioxide.
8. To do this plunge the plant material in boiling 90% alcohol.

WARNING: The alcohol must be boiled over a water bath.

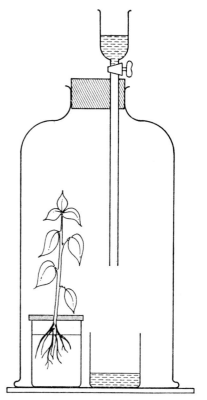

    The sample can then be kept in this state until required for the chromatogram.
    Each sample should be ground up in a mortar with pestle. Filter each sample.
    Evaporate the alcohol off **over a water bath.**
    Add 1 ml of water to the residue and swirl it around so that it includes all the residue.
    Spot this solution on to the chromatogram.
    Make other spots of sucrose, glucose and fructose.

9. Run the chromatogram in a phenol solvent. **Handle carefully.**
10. Develop the chromatogram in a 10% solution of resorcinol in acetone. Add a few drops of concentrated hydrochloric to the developing solution.
11. Yellow-brown spots will appear when the paper is dried.
12. Measure the activity of each part of the chromatogram by marking off the chromatogram in $\frac{1}{2}$ cm sections and determine the activity of each section with a Geiger counter with a limiting slit made from cardboard placed in front of the window. This slit should be 0·5 cm $\times$ 2·5 cm.

Results

Identify each of the sugars formed in the plant material and plot the activity of each section against its position on the chromatogram.

189

From these results what conclusions can you draw about the synthetic pathways of carbohydrates in photosynthesis. Compare the results you obtain from each of the chromatograms.

Requirements

Apparatus

Geiger-Muller tube, Mullard MX 168 or MX 168/01
Belljars
Ground glass plates
Small beakers
Dropping funnels
X-ray film. (Kodak Crystallex or Kodirex or Industrex type D)
Chromatography equipment

Reagents

$Na_2C^{14}O_3$. This can be obtained in tablet form from J. A. Radley (Laboratories) Ltd., 220 Elgar Road, Reading, Berkshire. The permitted amounts of radionuclides are dispensed into inactive tablets of appropriate chemicals. The tablet is dried and placed in a capsule. This carries full instructions and a description of the contents.
It is recommended that this is made into a solution as soon as it is received
2 M HCl, and concentrated HCl
Phenol solvent (28 g of phenol in 12 ml of water)
10% resorcinol in acetone
Sucrose, fructose and glucose

Biological

Pot plants

The catalytic properties of extracted chloroplasts (The Hill reaction)

The purpose of this experiment is to show that isolated chloroplasts can split the water molecule to release oxygen.

Chloroplast extracts, when illuminated have the power to reduce ferric ions and to liberate oxygen. They cannot use carbon dioxide. The hydrogen is used to reduce a hydrogen acceptor. A redox dye is used for this purpose in the experiment, namely 2:6 dichlorophenolindophenol.

$$2:6\,D + H_2O \rightarrow 2:6\,DH_2 + \tfrac{1}{2}O_2$$
$$\text{blue} \qquad\qquad\qquad \text{colourless}$$

In the plant photosynthetic system it seems likely that TPN or COII (coenzyme triphosphopyridine nucleotide) is the hydrogen acceptor. DPN may also be involved.

190

Procedure

1. Homogenise spinach in 0·5 M sucrose for 30 seconds, using a blender.
   All solutions should be kept ice-cold throughout the experiment.
2. Filter the homogenate through muslin and centrifuge at a low speed for 10 minutes. This throws down cell wall debris and starch.
3. Pour off the supernatant and re-centrifuge at high speed for 10 minutes. This throws down the chloroplasts.
   When centrifuging the large tube should contain a freezing mixture and the experimental material should be in a smaller tube in the centre of this.
4. Again pour off the supernatant and re-suspend the chloroplasts in 4 ml of 0·5 M sucrose.
5. Prepare 8 tubes as follows:

   Tubes 1 and 2. 2 ml 2 : 6 D (2.5×10^{-4}M)
   2 ml M/10 phosphate buffer, pH 6·5
   2 ml chloroplast suspension.
   Make up to 10 ml with water.

   Tubes 3 and 4. As for 1 and 2 but using boiled chloroplast suspension as a control.

   Tubes 5 and 6. As for 1 and 2 but adding 2 drops of potassium cyanide, M/1,000, to the chloroplast suspension.

   Tubes 7 and 8. As for 1 and 2 but without any dye.

WARNING: Do not use mouth pipette.

6. Illuminate tubes 1, 3, 5 and 7 brightly and leave the others in the dark.
7. Measure the time for the dye to bleach in the tubes in the light.

Questions

1. Why is potassium cyanide used in the experiment?
2. What can you infer about the nature of the hydrogen donors from the results you obtain in the cyanide treated tubes?

Requirements

Apparatus	Reagents
Blender or liquidiser	0·5 M sucrose
Muslin	Freezing mixture
Centrifuge	2 : 6 dichlorophenolindolphenol
8 test tubes	(2.5×10^{-4} M)
Strong lamps	M/10 phosphate buffer, pH 6·5
	KCN

Biological

Spinach

A simple method of measuring the respiratory rate of small animals

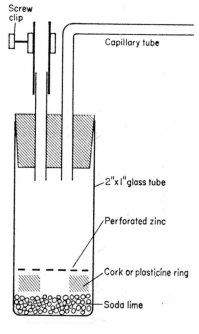

Screw clip

Capillary tube

2"x1"glass tube

Perforated zinc

Cork or plasticine ring

Soda lime

Construction of apparatus

This experiment requires the use of two simple respirometers made up as follows.

Take a 3 in × 1 in flat bottomed specimen tube and place in it a layer of self-indicating soda lime (e.g. 'Carbosorb' which is available from British Drug Houses Ltd). This type of soda lime contains an indicator which changes colour when the soda lime loses its capacity to absorb carbon dioxide. Above the soda lime place a disc of perforated zinc or wire gauze, supported by a ring of cork or plasticine.

The top of the tube is fitted with a two-holed rubber bung. Insert a piece of capillary tube, bent to a right angle in one hole and a short piece of glass tube connected to a small piece of pressure tubing which can be closed by means of a screw clip.

Procedure

1. Calibrate the respirometer by drawing a column of mercury into the capillary tube. Measure the length of the column of mercury and weigh it. The volume of the tube per centimetre length is found from the formula

$$\text{Volume of tube} = \frac{\text{Weight of mercury}}{\text{Length of column} \times 13 \cdot 6} \text{ ml per cm}$$

2. Weigh 5 woodlice.
3. Place the woodlice in one respirometer. The other is used as a control (thermobarometer).
4. Place both respirometer and the thermobarometer in a water bath so that they are both fully immersed to the level of the rubber bung.
5. Leave both screw clips open and allow the apparatus to equilibrate for 5 minutes.
6. Close the screw clip on both pieces of apparatus.
7. With a pipette, place a small drop of manometric fluid at the ends of the capillary tubes so that some is sucked in as oxygen is converted to carbon dioxide and absorbed.
8. Time the movement of the manometric fluid over a measured distance.
9. Repeat several times. If a disposable plastic hypodermic syringe is inserted in the pressure tubing the manometric fluid can be readily returned to the end of the capillary tube. (Plastic syringes are available from all chemists shops at 7–9d. each.)
10. Calculate the oxygen consumption per hour from the formula:

$$O_2 \text{ consumed} = \frac{\text{Vol of } O_2 \text{ used} \times 60}{\text{Time in min} \times \text{Weight of animals}} \text{ ml/g/hour}$$

If accurate comparisons are to be made this figure must be converted to NTP (use air temperature at level of capillary—not bath temperature).

Note. If any movement of the manometer fluid in the thermo-barometer occurs during the experiment, it must be added to, or subtracted from, the figures obtained in the respirometer.

Effect of temperature upon oxygen consumption

Procedure
1. Use the procedure outlined above to measure oxygen consumption with the respirometer in a water bath set at 20°C. (Unless ambient is higher.)
2. Adjust water bath to 25°C, allow 5 minutes to equilibrate before taking readings.
3. Repeat experiment with the temperature raised in steps of 5 degC until 40°C is reached.
4. If a refrigerator is available the effects of sub-ambient temperatures can be measured.
5. Plot a graph of oxygen consumption against temperature.

Requirements

Apparatus

2 simple respirometers
Water bath
Retort stand and clamp (optional)
Thermometer
Balance

Reagents

Soda lime (preferably self-indicating)

Biological

Woodlice

Determination of the respiratory quotient of germinating peas

The ratio of carbon dioxide evolved to oxygen consumed, $\dfrac{CO_2}{O_2}$, is called the respiratory quotient or R.Q.

193

When the R.Q. is measured experimentally it is found to vary according to the nature of the substance being respired. The R.Q. for the aerobic respiration of sugar is approximately 1·0. If fat is the respiratory substrate, the amount of oxygen consumed is greater and the R.Q. value may fall as low as 0·7. In some plant material, the R.Q. value may fall as low as 0·1 due to the incomplete oxidation of sugar to form malic or oxalic acid instead of carbon dioxide. Estimation of the R.Q. value for plant material must be carried out in the dark to prevent any of the carbon dioxide produced being used in photosynthesis. If protein forms part of the substrate, the R.Q. value will be in the region of 0·8 to 0·9.

Respiration involves a number of complex chemical changes, but the overall reaction may be summarised for the respiration of carbohydrate and fat thus:

Carbohydrate

$$C_6H_{12}O_6 + 6O_2 \rightarrow 6CO_2 + 6H_2O$$

as 6 molecules of O_2 are consumed and 6 molecules of CO_2 are produced, the respiratory quotient is given thus:

$$R.Q. = \frac{6}{6} = 1\cdot0$$

Fat (glycerol tristearate)

$$CH_2O.CO.C_{17}H_{35}$$
$$|$$
$$2\ CHO.CO.C_{17}H_{35} + 163\ O_2 \rightarrow 114\ CO_2 + 110\ H_2O$$
$$|$$
$$CH_2O.CO.C_{17}H_{35}$$

as 163 molecules of O_2 and 114 molecules of CO_2 are involved, the R.Q. value $= \dfrac{114}{163} = 0\cdot7$.

Procedure

The apparatus illustrated in the diagram is used to measure oxygen uptake and carbon dioxide output simultaneously.

A length of capillary tubing is attached to the side arm of the flask. This tube should be bent so that it can hang vertically.

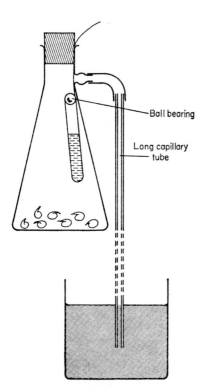

Ball bearing

Long capillary tube

1. Place 15 g of germinating peas in the flask.
2. Place 10 ml of 10 N sodium hydroxide in a small tube and stopper this with a ball-bearing. It is important that this ball-bearing does not come in contact with the sodium hydroxide, and that it completely closes the neck of the tube.
3. Place the tube in the flask with the aid of a piece of cotton tied round its neck. Drop the cotton into the flask.
4. Stopper the flask. It is important that the stopper and the side-arm connection are air-tight.
5. Place the other end of the capillary tube under the surface of water dyed with methylene blue or eosin, and containing Teepol.

6. The flask should be placed in a water bath kept at a constant temperature.

7. **If there is any likelihood of photosynthesis taking place the experiment should be run in the dark room,** or the flask should be covered in black paper.

8. Allow the peas to respire for $1\frac{1}{2}$ hours.

9. At the end of this time mark the height of the dye in the capillary tube with a small strip of Sellotape.
   Also mark the level of the liquid in the beaker on the capillary tube.

10. Now taking care to keep the end of the capillary tube below the surface of the dye solution, use a magnet to remove the ball-bearing from the small tube in the flask.

11. Gently tilt the flask to spill the sodium hydroxide into the flask. Agitate the flask slightly for a few moments, still taking care not to remove the end of the capillary tube from the dye solution.

    Do not warm the flask with your hand.

12. As the carbon dioxide is absorbed by the sodium hydroxide, the level of solution in the capillary tube will rise. Mark this new level when it has ceased rising.

13. Remove the side-arm.

14. To measure the volume of liquid in the side-arm at each stage of the experiment, partly fill a finely graduated burette with some of the dye solution and note the volume.

15. Pipette some of the dye solution into the side-arm up to the final mark.
    Release this solution into the burette until only the solution below the mark corresponding to the level of solution in the beaker of dye, remains in the tube.
    Note the volume in the burette, and hence the volume of solution which had been in the tube.

16. Repeat this procedure for the other mark on the side-arm.

17. Measure the volume of the apparatus. This can be taken as being the volume of the flask. To do this plug the side-arm and fill it with water and stopper as in the experiment. Then measure the volume of water in a measuring cylinder.

18. Similarly measure the volume of the peas, and the volume of the sodium hydroxide tube with the ball-bearing in place.

Results

Let D_1 = the distance of solution level in the side-arm tube above the level in the beaker before NaOH was added.

V_1 = the volume of solution in the side-arm tube just before NaOH was added.

D_2 = the distance of solution level in side-arm tube above the level in the beaker after the addition of NaOH.

195

V_2 = the volume of solution in the side-arm tube after the addition of NaOH.

V_t = the total volume of the apparatus, excluding peas and NaOH tube.

V_1 is a measure of the difference between O_2 uptake and CO_2 production.

V_2 is a measure of the absolute uptake of O_2 as all the CO_2 will have been removed by the NaOH.

Therefore the absolute volume of CO_2 produced $= V_2 - V_1$, i.e. neglecting the CO_2 content of the atmosphere.

Before making the calculation, the volumes V_1 and V_2 must be corrected to the volumes they would occupy at atmospheric pressure.

Now the flask volume at atmospheric pressure

$$= V_t \times \frac{\text{flask pressure}}{\text{atmospheric pressure}}$$

Pressure in flask when V_1 was read will be atmospheric pressure—weight of water in side-arm above level in beaker.

To express this in millimetres of mercury, divide D_1 by 13.

Represent $D_1/13$ by P_1.

Flask pressure now equals $760 - P_1$.

Total volume V in the apparatus when V_1 is read

$$= \frac{(V_t - V_1)\,(760 - P_1)}{760}$$

Corrected $V_{1'} = V_t - V$, i.e. the true net volume change.

Then determine $V_{2'}$, the true total O_2 uptake, with a similar calculation using V_2, D_2 and P_2.

$$\text{R.Q.} = \frac{V_{2'} - V_{1'}}{V_{2'}}$$

R.Q. values and possible substrates: (after Strafford, 1965)

R.Q.	Substrate
>1·0	Carbohydrate with some anaerobic respiration
	Carbohydrate synthesised from organic acids
	Organic acids
1·0	Carbohydrates
0·99	Proteins with ammonia formation
0·8	Proteins with amide formation
0·7	Fats
0·5	Fats with associated carbohydrate synthesis
0·3	Carbohydrates with associated organic acid synthesis

In interpreting the results remember that two substrates may be used simultaneously.

We have ignored the different solubilities of the two gases as they will not be important when using such crude apparatus as this.

Requirements

Apparatus

Flask with a side-arm, stopper, and rubber tube to attach it to capillary tube. Capillary tube with right angle near one end; about 24 inches long

Beaker

Small glass tube which will easily fit inside the flask

Ball-bearing which fits in neck of this tube to make an air-tight seal

Black paper or a dark room

Water bath set at any convenient temperature (around room temperature) but it must be constant

Finely graduated burette

Reagents

10 N NaOH

Methylene blue in water with a little detergent added such as Teepol or use a solution of 23 g of NaCl, 5 g of sodium choleate and 100 mg of Evan blue in water and dilute to 500 ml

Biological

Germinating peas

Measurement of the respiratory rate of small animals using the Dixon-Barcroft apparatus

Many types of apparatus can be used to measure respiration manometrically. The Dixon-Barcroft apparatus is of the constant volume type, i.e. to take a reading, the internal volume of the apparatus is adjusted to the value at the outset of the experiment. This means that the gas in the apparatus is at the same pressure for each reading, and subsequent readings are directly comparable with each other.

Apparatus

The apparatus consists essentially of two flasks attached to the limbs of a manometer. A third limb, graduated directly in ml and filled with mercury is placed between one of the flasks and the manometer. The level of the mercury may be adjusted by the movement of a screw clamp. The two flasks can be opened to the atmosphere by the taps T_1 and T_2. Each flask has a central cell to hold potassium hydroxide solution and a wick of rolled filter paper. Animals are placed in one flask F_1 and the second flask F_2 acts as a control (thermobarometer).

Make sure that you understand the apparatus before using it. Opening the taps in the wrong sequence may bring about mixture of the manometer fluid and the mercury. Cleaning and refilling the apparatus is very time consuming.

197

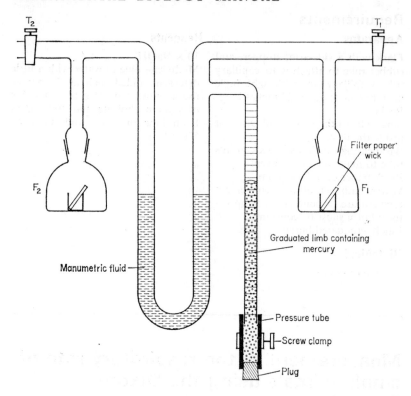

Procedure

1. Open the taps T_1 and T_2.
2. Roll up two pieces of filter paper, approximately $1.5 \, \text{cm} \times 3 \, \text{cm}$ and use forceps to insert them on the central cells of the flasks to form wicks.
3. Place $0.3 \, \text{ml}$ 10% potassium hydroxide solution in each cell.
4. Carefully weigh 6–10 animals (woodlice, mealworms, centipedes or blowfly larvae), and place them in the flask F_1.
5. Lower the flasks, so that they are completely submerged, into a thermostatically controlled water bath. Leave for 10 minutes to equilibrate.
6. Close the taps T_1 and T_2. Record the level of the mercury in the graduated arm.
7. If no movement of the manometer fluid occurs, or the level changes and then falls back, there is an air leak in the apparatus and all ground glass joints should be checked for tightness of fit and greased with burette grease if necessary.
8. After 10 minutes take a reading. Adjust the screw clip on the graduated arm so that the manometer fluids are returned to the same level in each arm of the manometer. Read the new level of the mercury.
9. Repeat to obtain a number of readings.
10. Calculate the average volume of oxygen consumed in 10 minutes. Multiply by 6 to obtain volume of oxygen consumed

in 1 hour. Divide by the weight of the animals used to obtain the ml/g/hr of oxygen consumed.

Temperature

Carry out the experiment at a range of temperatures between 20°C (or ambient if this is higher), and 40°C. If more than one apparatus is available, different groups of the class may obtain data for different temperatures. Plot a graph to show the relationship between temperature and oxygen consumption.

N.B. If an accurate comparison of data obtained on different occasions is desired, the volume of oxygen consumed must be converted to NTP.

When using a thermostatically controlled water bath it is essential to check the actual water temperature with a thermometer as dial settings are only approximate. The flask will only assume the temperature of the bath if it is completely immersed.

Projects

1. Compare the oxygen consumption of different animals.
2. Compare the oxygen consumption of larval, pupal, and adult blowflies.

Requirements

Apparatus

Dixon-Barcroft apparatus
Thermostatically controlled water bath
Thermometer

Reagents

10% potassium hydroxide solution

Biological material

Woodlice, mealworms, centipedes or blowfly larvae

Calculation of the respiratory quotient (R.Q.) using the Dixon-Barcroft apparatus

Procedure

1. Set up the Dixon-Barcroft apparatus in a water bath as described in the previous experiment but **omit the 10% potassium hydroxide solution.**
2. Weigh 6–10 animals (woodlice, mealworms, centipedes or blowfly larvae) and place them in the flask F_1.
3. Leave the apparatus and animals for 10 minutes to equilibrate with the bath set at 25°C.
4. Close the taps and allow the experiment to run for exactly 30 minutes. **Keep a careful watch on the level of the manometer**

fluid and adjust its level by using the screw clamp on the arm containing mercury if necessary.

5. After 30 minutes record the change in the volume of the gas in the apparatus after adjusting the screw clamp to bring the fluid in the manometer arms level.
6. Remove the apparatus from the water bath and place 0·3 ml of 10% potassium hydroxide solution in the central cell of each flask.
7. Return the apparatus to the water bath, allow 10 minutes to equilibrate and then close the taps.
8. After 30 minutes record the change in volume of the gas in the apparatus.
9. By comparison of the figures obtained in steps 5 and 8, we can obtain the volume of oxygen consumed and the volume of carbon dioxide produced.

$$R.Q. = \frac{CO_2 \text{ produced (Volume at N.T.P.)}}{O_2 \text{ consumed (Volume at N.T.P.)}}$$

Further work

The Dixon-Barcroft apparatus may also be used to measure the R.Q. of germinating seeds.

This experiment can be carried out using the simple respirometer described previously but will be found to be rather more tedious and less accurate.

Requirements

Apparatus

Balance
Dixon-Barcroft apparatus
Thermostatically controlled water bath
Thermometer

Biological material

Woodlice, mealworms, centipedes or blowfly larvae

Reagents

10% potassium hydroxide solution

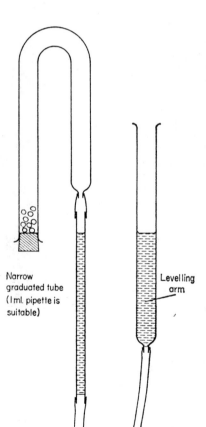

Narrow
graduated tube
(1ml. pipette is
suitable)

Levelling
arm

Determination of the respiratory quotient of germinating seeds using the Ganong respirometer

Procedure

1. Two sets of apparatus are available for this experiment. One is rather expensive, consisting of a bulb in which the seeds are placed, and a stopper with a lateral hole which can be closed

to begin the experiment. The other merely consists of an inverted U-tube with a rubber stopper.

Each type is connected to a graduated cylinder and a levelling cylinder.

Two sets of equipment are required for each experiment.

2. Fill one with a 10% solution of caustic potash, and the material. If using the U-tube place about 5 g of germinating seeds in the U-tube; if using the other type of apparatus, place 2 ml of material in the small bulb at the base of the main bulb.

This must be accurately measured as the apparatus is made such that the bulb and the measuring cylinder contains a volume of 102 ml down to the bottom graduation mark.

3. Adjust the level of the potash such that the levelling cylinder is nearly full and the potash is just up to the 100 ml mark on the graduated cylinder. In the other type of equipment it must be near the bottom of the graduated tube. The levels must be the same.

4. In the U-tube equipment, the experiment begins once the apparatus is complete. Note the time and allow the experiment to run for 24 hours, noting also the levels and temperatures at the beginning and end of the period.

In the bulb apparatus, there is a lateral hole in the stopper which can be closed or opened and the experiment only commences when this is closed.

5. Set up a second set of apparatus containing the material under investigation and water.

In this case after levelling, the levelling cylinder must be about half full and the water level in the measuring cylinder must be about half way up it.

6. Again read the levels, and note the temperature at the beginning and end of the 24 hour period.

Results

The potash will absorb the CO_2 and so the amount of O_2 consumed will be given by the rise in level of the apparatus containing potash.

The amount of carbon dioxide produced is found by comparing the levels of both sets of apparatus.

If more CO_2 is produced than O_2 used, the level of water will go down and this new volume must be added to the O_2 volume to give the volume of CO_2.

If there is a reduction in volume, then the amount of CO_2 produced is found by subtracting this reduction in volume from the O_2 volume.

If there is no change in volume, the amounts of CO_2 and O_2 used must be the same.

$$R.Q. = \frac{\text{volume of } CO_2 \text{ produced}}{\text{volume of } O_2 \text{ used}}$$

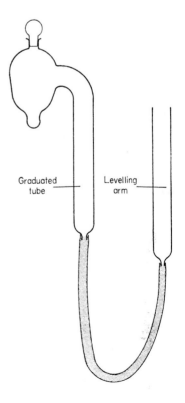

Graduated tube

Levelling arm

Questions

1. Why have you been told to measure the temperature before and after the experiment?
2. A third piece of apparatus could be set with dead peas and water; What would the reason for this be?
   In this case in addition to killing the peas, in what other way would you treat them?

Requirements

Apparatus

2 or 3 sets of equipment as shown in the diagrams

Reagents

10% potash solution

Biological

Germinating seeds or yeast

Determination of the rate of respiration using a Pettenkoffer tube

This technique is used to measure the rate of respiration by determining the amount of carbon dioxide produced by a known weight or volume of material in a known time.

Carbon dioxide free air is supplied to the respiring material, and the gases formed are passed through barium hydroxide in which the carbon dioxide reacts to form barium carbonate. This is insoluble. At the end of the experimental period the unchanged barium hydroxide is titrated against a standard hydrochloric acid. This measures the amount of barium hydroxide left and hence the amount which has reacted with the carbon dioxide.

This technique can be used to determine the Q_{10} value of respiring material.

Procedure

1. Attach three flasks to the Pettenkoffer tube as shown in the diagram.
2. A gas current can be passed through the apparatus either by attaching a filter pump to the bulb end of the Pettenkoffer tube or by attaching an air pump to the first flask.
3. Fill the flasks as shown in the diagram.
   The respiring material can be any actively respiring biological material. 2 g of yeast per 30 ml of a 5% sucrose solution is an ideal material. Germinating seeds are also suitable.
   This flask can be placed in water baths at different tempera-

tures so that the Q_{10} value for this reaction can be determined. If a belljar is used the amount of carbon dioxide produced by whole pot plants can be determined.

4. Place 60 ml of the barium hydroxide solution in the Pettenkoffer tube.
5. It is convenient to fit a two way tap so that the tube can be readily changed.
6. In setting up the apparatus it is essential that it is air-tight throughout. Dreschel bottles with ground glass stoppers are useful for this purpose.
7. Regulate the rate of flow of the gas so that it is about 60 bubbles per minute through the Pettenkoffer tube.
8. Discard this first tube by switching the two way tap over to the second tube and start the timing of the experiment from this moment.
9. Run the experiment for one hour.
10. Immediately a tube is removed from the apparatus, pour the contents into a flask and wash the tube with boiled de-ionised water. Add the washings to the flask.
Titrate 10 ml of this barium hydroxide solution against N/10 hydrochloric acid.
Use phenolphthalein as the indicator.

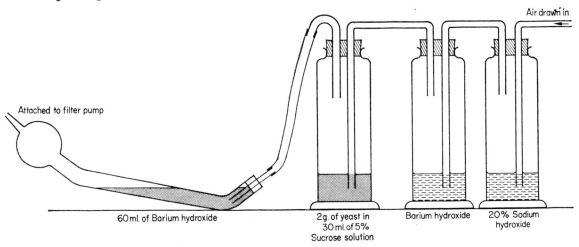

Air drawn in

Attached to filter pump

60 ml. of Barium hydroxide

2 g. of yeast in
30 ml. of 5%
Sucrose solution

Barium hydroxide

20% Sodium
hydroxide

11. Determine the strength of the stock barium hydroxide by also titrating this against the N/10 hydrochloric acid.
Although N/10 barium hydroxide should be used, its strength must be accurately determined.

Results

From the equations for the reactions of barium hydroxide with (a) HCl, and (b) CO_2, it can be determined that 73 g of HCl is equivalent to 44 g of CO_2.
Therefore 1 litre of N HCl contains 36·5 g and therefore is equivalent to 22 g of CO_2.

If x represents the amount of $N/10$ HCl needed to neutralise the stock solution of barium hydroxide, and y the amount needed to neutralise the barium hydroxide from the tube, the weight of CO_2 evolved $= (2 \cdot 2/1,000 \times x) - (2 \cdot 2/1,000 \times y)$ g.

This is in 10 ml. Multiply this figure by 6 to determine the amount in the 60 ml of barium hydroxide.

Similar calculations can be performed to find the weight evolved at different temperatures and hence Q_{10} values can be measured.

Further work

The experiment can be performed under anaerobic conditions and the amount of CO_2 produced can be compared with that under aerobic conditions.

To do this pass nitrogen gas through the tube from a cylinder.

Are Q_{10} values under anaerobic conditions any different to those under aerobic conditions?

Requirements

Apparatus

2 Pettenkoffer tubes
3 flasks with tight fitting stoppers
Glass tubes as shown in the diagram
Belljar
Filter pump or an air pump
2 way tap

Reagents

$N/10$ barium hydroxide
$N/10$ HCl
5% sucrose solution
20% NaOH
Phenolphthalein

Biological

Yeast
Germinating seeds
Pot plants

The effects of temperature on the heart beat of 'Daphnia'

The rate of a physiological activity increases as the temperature rises until a point is reached at which injury to the tissues results; then the activity decreases and finally death follows. Within a range of about 10 deg C above and below the average temperature of the animal, the rate of a process is usually doubled for every 10 deg C rise in temperature. Sudden changes in temperature may produce strange results due to the shock effects, photochemical reactions are affected to lesser extent also.

The effect of temperature is normally expressed in terms of the **temperature coefficient** (Q_{10}). This is the ratio of the rate of the activity at one temperature to the rate of the reaction at a temperature of 10 deg C lower.

$$Q_{10} = \frac{T + 10}{T}$$

Procedure

1. Allow the *Daphnia* culture to come to room temperature.
2. With a pipette, transfer one *Daphnia* to a cavity slide.
3. Soak up the excess liquid with a piece of filter paper so that the animal will lie on its side to make the heart more readily visible.

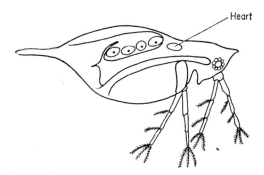

4. The heart beat will probably be so rapid that the following method will be necessary to keep a record. One student should measure 20 seconds with a clock, shouting start and stop at the beginning and end of the period. The other student should mark a dot on a piece of paper each time the heart contracts. Make at least three counts and determine the average.
5. Place *Daphnia* in a 50 ml beaker of culture medium, and alter its temperature by placing it in a 250 ml beaker of warm water or ice.
6. Determine the rate of beat at 5, 10, 15, 20, 25, 30 and 35°C. The temperature of the *Daphnia* medium will soon fall when it is placed on the cavity slide but you will have to assume that this effect will be similar in each case.

  Between each measurement at each temperature, it will probably be necessary to place the animal back in the temperature bath for a few minutes.

  When placing the animal in a new temperature, allow it 5 minutes at that temperature before determining the heart beat rate.

Note. An individual may try this experiment over a limited temperature range using one animal if possible, or it may be performed as a class experiment. In this case several *Daphnia* will have to be used.

Results

Plot a graph of the mean frequency of the heart beat in beats per minute against temperature.

Draw the best curve through these points.

Calculate the standard error for each set of readings.

In comparing the means of two sets of data (i.e. the means at two different temperatures) we want to know whether we can attach any significance to the differences in the mean values. The difference between two sets of means can be tested for significance if the standard error of the difference is calculated. This takes into account the variability of the data (for statistical details see the appendix, page 295).

Apply the t-test to the results to test for their significance.

Calculate the Q_{10} for the ranges 5–15, 10–20, 15–25, 20–30, and 25–35°C.

Requirements

Apparatus

Binocular microscope
Cavity slide and coverslip
50 ml beakers, that is one for each water bath
7 250 ml beakers to act as water baths, one at each of the following temperatures: 5°, 10°, 15°, 20°, 25°, 30° and 35°C
Clock

Reagents

Ice

Biological

Daphnia

An investigation of the response of the respiratory control centre to a change in carbon dioxide concentration

The rate of pulmonary ventilation is controlled by the respiratory centre which responds to changes in the CO_2 concentration of the blood. If the CO_2 concentration rises both depth, and rate of lung ventilation increase. A decrease in the CO_2 concentration of the blood is followed by a rapid fall in the ventilation rate.

Procedure

1. Set up the kymograph apparatus to revolve at 2·5 or 1·25 cm per second.
2. A stethograph (available from Messrs. Palmers), is connected to a tambour with a piece of pressure tubing. The tambour operates a writing lever.
3. The stethograph is placed around the chest of a member of the class. It is best held in place by string 'braces' over the shoulder,

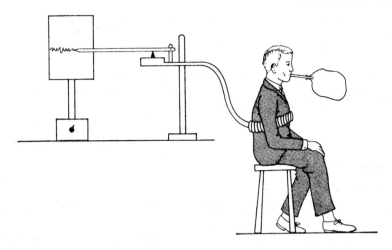

thus preventing the stethograph from slipping down the chest during the recording.

4. Record a short trace of normal respiration.
5. Give the subject a polythene bag, Sellotaped to a piece of large diameter tubing to form a mouthpiece. The subject breathes in and out of the bag, thus increasing the alveolar CO_2, thereby making it more difficult to pass blood CO_2 into the lungs.
6. When the subject is ventilating rapidly, remove the bag so that he returns to normal air breathing.
7. With a mounted needle, mark an arrow on the trace at the time re-inhalation and restitution of normal breathing occurred.
8. Finish recording when a normal trace reappears.
9. Varnish the recording.

Questions

1. What changes in rate occurred? How would you explain them?
2. Did the depth of inhalation, exhalation, or both change?
3. Is this a valid experiment to isolate a single factor?

Requirements

Apparatus

Kymograph and paper
Smoking equipment
Stethograph
Tambour
Pressure tubing
Varnishing tray
Polythene bag fitted with mouth tube

Reagents

Kymograph varnish

An investigation of the excretory process in locusts

Excretory products in the haemocoel of an insect are actively passed through the walls of the Malpighian tubules into the lumen of the gut. The excretory principle may be demonstrated by placing dyes in the haemocoel and observing their subsequent removal and appearance in the lumen of the Malpighian tubules.

Procedure

Method A

1. Chill an adult locust in the bottom of a refrigerator or anaesthetise it for 10 minutes in carbon dioxide.
2. Make up a solution of 1% dye in locust Ringer. Suitable dyes in order of preference are: indigo-carmine; methylene blue; phenol red or neutral red.
3. Inject a few drops of the dye into the abdomen of the locust, using a finely drawn out glass pipette or a hypodermic syringe. Make the injection slightly to one side of the mid-line.
4. Allow the locust to recover in a warm room.
5. After 15 minutes the locust should be killed (ether or chloroform) and a dissection made to display the gut and Malpighian tubules. The gut and tubules should be carefully removed and examined under a low power microscope (mount in Ringer).
6. It will be seen that the Malpighian tubules are darkly stained. If some of the tubules are then examined under the × 40 objective, it will be observed that the bulk of the pigment has been concentrated in the lumen of the tubules.

Method B

1. Anaesthetise an adult locust by placing it in carbon dioxide for 10 minutes.
2. Open up the abdomen dorsally and place a few drops of 1% dye in Ringer in the cavity which surrounds the Malpighian tubules.
3. Observe the active movements of the tubules with a hand lens or binocular dissection microscope, as they pick up the dye.
4. The locust should be killed by decapitation before it recovers from the effects of the anaesthetic.
5. The tubules may be examined under the high power lens.

Requirements

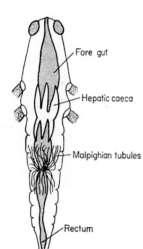

Fore gut

Hepatic caeca

Malpighian tubules

Rectum

Apparatus

Hand lens or binocular microscope
Syringe or finely-drawn glass pipette
Dissection instruments
Microscope and slides

Biological material

Adult locust

Reagents

Locust Ringer
Indigo-carmine
Methylene blue
Phenol red or neutral red

208

Kymograph recording

A number of physiological experiments demand a record which shows, as accurately as possible, the frequency, intensity and duration of some activity. Such records may be made on very expensive ink recorders or on smoked paper, mounted on the drum of a kymograph apparatus.

The paper used on the kymograph has a very smooth, shiny surface so that it offers little frictional resistance to the recording point. The paper is placed around the drum and its edges stuck together, care being taken that the paper edges overlap in the same direction that the recording point travels. The drum is then smoked by rotating it slowly over a smoky flame. If proprietary smoking equipment is not available, coal gas can be passed through benzene in a flask and burnt in a bunsen burner fitted with a spreader or fishtail. The airhole of the bunsen burner should be closed.

Recording points

A useful record will probably require the simultaneous recording of several pieces of information, each requiring a separate recording point. The traces usually needed are:

1. The physiological response.
2. The point at which a stimulus is applied and its duration.
3. Some accurate time record.

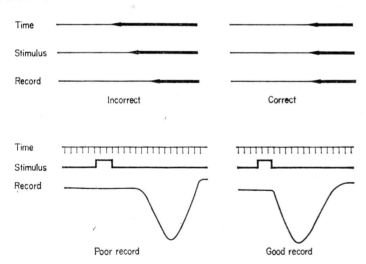

Care should be taken to observe the following details when setting up the recording points:

   (a) All the recording points should lie in a vertical line.
   (b) Place the stimulus and time marker points at the top of the drum, with the response lever below.
   (c) Ensure that the response lever is positioned so that all of its

209

movement can be recorded on the drum without crossing the traces of the time and stimulus markers.

(*d*) The response lever point must be placed so that it lightly touches the smoked paper—it must give a clean, continuous line with the minimum of friction. Any unnecessary friction will restrict natural movement of the organ under observation.

(*e*) With the drum stationary, stimulate the organ and check that the vertical movement of the lever is parallel to the drum, so that a complete trace is obtained at an even pressure.

(*f*) The recording levers should be horizontal when they are in the neutral position, i.e. before a stimulus is applied.

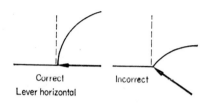

Correct — Lever horizontal · Incorrect

Types of response lever

Isotonic lever

In this type of lever the organ under investigation is kept under a constant tension by the use of a weight. Isotonic levers record the amount and frequency of contraction but not the tension. They are used in making records of smooth muscle and heart muscle.

Isometric levers

This type of lever has a fairly strong spring in place of the weight used in the isotonic lever. Such an arrangement enables the experimenter to record changes in muscle tension as the movement of the recording lever is proportional to the tension which the muscle places on the spring. Used in examining the responses of voluntary muscle.

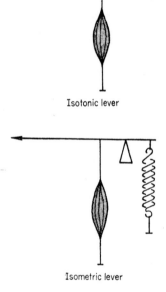

Isotonic lever

Isometric lever

Tambours

These are used to apply changes in volume to a recording lever. They can be attached to the body by means of pressure tubing. Tambours must be used in conjunction with a suitable device for picking up a response from the body, e.g. a stethograph is used to record respiratory rate and depth of breathing or a receiving cup is used to record arterial pulse rate.

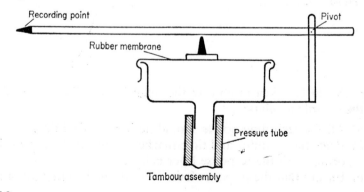

Recording point · Pivot · Rubber membrane · Pressure tube · Tambour assembly

Stimulus marker

In nerve muscle preparations a marking point attached to an electro magnetic make and break, is used to mark the duration and point of application of a stimulus.

Time marker

A very wide range of time marking equipment is available for use with kymograph apparatus, much of it being more sophisticated —and expensive, than is necessary for school work. A satisfactory marker for school use may be made by Sellotaping a celluloid writing point (cut from 35 mm, 16 mm or 8 mm film) to a tuning fork with a wavelength of 100 cycles per second. The tuning fork is struck on the back of the hand and the celluloid point applied to the revolving drum.

An investigation of a nerve-muscle preparation, using the sciatic nerve and gastrocnemius muscle of a frog

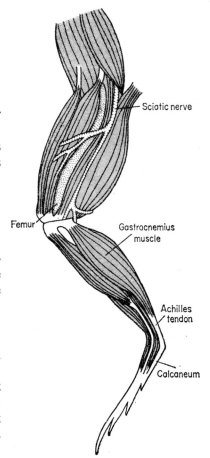

Aims

1. To show that an electrical impulse applied to a nerve may stimulate muscle contraction.
2. To show that groups of muscle fibres are innervated by axons which only pass an impulse if they are excited by a stimulus which exceeds their threshold value.
3. To show summation.
4. To show tetanus.

Procedure

Read the notes on kymograph recording in the kymograph technique section of this book. Prepare the smoked paper and arrange all the kymograph equipment before making the nerve-muscle dissection.

Gastrocnemius preparation

1. Kill the frog by pithing. This must be done by the teacher in the preparation room.
2. Hold the frog firmly by the anterior end in one hand. Cut the loose skin around the middle of the frog with scissors. Grip the skin firmly with the thumb and first finger and exert a strong decided pull so that the skin is stripped from the legs in one movement.

211

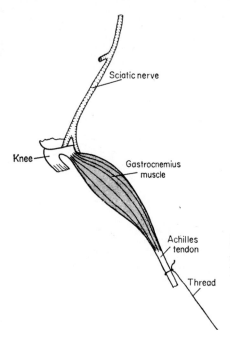

3. Open the abdomen and remove the abdominal viscera without touching the underlying nerves.
4. Cut the whole animal across just in front of the iliac attachment, using strong scissors.
5. Still using strong scissors, cut down the spine and through the pelvic girdle to bisect the material. Each part will yield a preparation.
6. Using the fingers, part the muscles on the back of the leg to expose the sciatic nerve from the pelvis to the knee.
7. Remove all muscles above the knee. As far as possible avoid touching the nerve with metal instruments.
8. Cut the femur just above the knee.
9. Tie a thread around the Achilles tendon and cut through the tendon where it attaches to the calcaneum.
10. Use the thread to tear the gastrocnemius muscle away from the tibia, then cut the tibia below the knee.
11. Mount the preparation on a cork board, pinning it through the knee and laying the sciatic nerve across the electrodes. Attach the thread to the isotonic lever of the kymograph. Keep the preparation moist with frog Ringer at all times.

Investigation of a maximal contraction

When a nerve impulse, large enough to excite all components of a motor system reaches a muscle, the muscle will undergo a maximal contraction. The kymograph record will show that there is an interval between the application of the stimulus and the onset of the contraction. This is known as the latent period. The latent

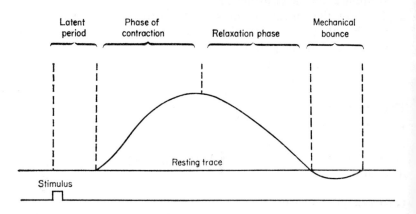

period is followed by a phase of contraction which gives way to a phase of relaxation. After the relaxation phase there may be a tendency for the trace to drop slightly below the resting level; this is mechanical bounce due to the weight of the lever.

1. Set up the gastrocnemius preparation and the kymograph as shown in the diagram.

212

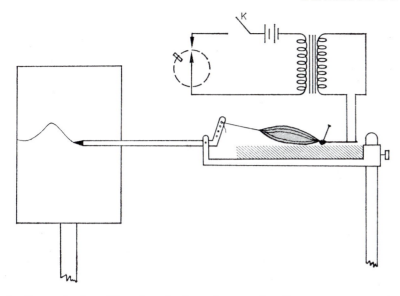

2. Open the key K to break the primary circuit.
3. Set the drum speed to 5 cm per second and move the recording lever so that it just makes contact with the drum to give a resting muscle line.
4. Close the key to complete the primary circuit, with the induction coil set to give a maximum stimulus, record one twitch and switch off.
5. With the key still closed, rotate the drum very slowly by hand so that a twitch is recorded as a vertical line at the stimulus point. (If a stimulus marker is available it may be wired into the circuit so that the stimulus and twitch can be recorded simultaneously.)
6. Remove the recording lever from the drum, switch on and record a time trace with a 100 cycle tuning fork.
7. Cut the paper from the drum with a sharp scalpel. (**Cut outwards so that the surface of the drum is not scored.**)
8. With the point of a needle, record any information required on the smoked paper. Dip the kymograph paper in a quick-drying varnish (see appendix) to make the record permanent.

Investigation of sub-maximal contractions

The muscle is made up of a number of motor units. Each unit is composed of a neuron and a number of muscle fibres associated with it. The axons in a nerve have different thresholds and only pass an impulse if the stimulus applied exceeds their threshold. If the threshold value of a particular neuron is reached, the muscle fibres which it controls contract fully. (It is an **'all or nothing' response.)**

1. Set up the gastrocnemius preparation on the kymograph.
2. Move the coils of the induction coil apart to their maximum distance to give a very weak stimulus.

3. Set the drum revolving at 5 cm per second.
4. Move the coils closer by small steps until a minimal twitch is seen.
5. Increase the stimulus slowly until a maximal twitch is obtained.

By how many steps is the maximal twitch reached? Is there a steady increase in the size of the response which cannot be recognised as separate steps?

Investigation of the effect of increasing the rate of stimulation

If two stimuli are applied to the nerve-muscle preparation with a considerable interval of time between them, two separate muscle contractions of the same magnitude occur. If the two stimuli are only one or two milliseconds apart, the second will arrive during the **refractory period** of the first and have no effect. If the second stimulus arrives some time after the refractory period of the first but before the muscle has fully relaxed, a second twitch of greater magnitude than the first occurs. This is known as **summation.** If the rate of stimulation is increased slowly after summation has appeared, a point will be reached where a sustained, smooth contraction occurs. This sustained contraction is called a **tetanus.**

1. Set up the preparation as in the previous experiments.
2. Set the two drum contacts at 60° to each other. Switch on and close the key.
3. Switch off and move the drum contacts closer for each revolution until summation and tetanus are observed. Tetanus is most easily achieved and maintained if the drum contacts are disconnected and replaced by an adjustable make and break vibrator such as an electric bell. The induction coil supplied by Palmers for their kymograph has a built-in vibrator.
4. Use a 100 cycle tuning fork to calibrate your trace.

   1. At what time interval does summation begin?
   2. At what time interval does tetanus begin?
   3. What is the nature of the trace at stimulus frequencies which are higher than those needed to produce summation but lower than those required to produce a smooth tetanus?

Investigation of repeated stimulation

1. Set up the apparatus as in the first experiment.
2. Set the drum contacts to give a stimulus on each revolution. Allow the kymograph to run until the muscle response is almost extinct. Do not add Ringer during the experiment.

Investigation of different stimuli

1. Set up the preparation as in the first experiment, but short out the drum contact. Use a key to make or break the circuit. What happens on 'make' and on 'break'?
2. Remove the electrodes from the nerve. Does the preparation react if stimulated by (i) pinching the nerve end with forceps,

214

(ii) touching the nerve end with a hot needle, (iii) placing salt crystals, adrenaline solution or acetylcholine solution on the nerve end?

Requirements

Apparatus

Kymograph apparatus
Smoking equipment
Time marker or 100 cycle tuning fork
Stimulus marker (optional)
Varnish tray (dipping tray from Shandon chromatography apparatus is ideal)
Make and break key (Morse key will do)
Palmer induction coil *or* neon stimulator

Copper wire for completing the circuit
Cork board
Isotonic lever and stand
Pins
Dissection instruments
Cotton wool (for applying Ringer to the preparation)
Cotton
Two or three 2 volt accumulators

Reagents

Benzene
Frog Ringer solution
Sodium chloride crystals
Acetylcholine solution (1 in 10^{-6})
Adrenaline (1 in 10^{-7})

Biological

Frog

An experiment to measure the speed of conduction of a nerve impulse

The nerve impulse is a locally propagated wave of change in the electrical potential of the nerve membrane. An electrical stimulus of any magnitude above a threshold value, will initiate the passage of an impulse. If the distance between the stimulating electrodes and the recording electrodes is measured, and the time lag between application of the stimulus and the passage of the impulse past the recording electrodes is measured with suitable equipment, the rate of conduction in the nerve may be calculated. As it is doubtful that schools will possess the pulse generator and twin beam scope needed for this experiment we will consider a second, less sophisticated method of making the measurement. When a nerve impulse passes along the sciatic nerve it causes contraction of the gastrocnemius muscle. The time when the electrical stimulus is applied is recorded on a kymograph trace of the muscle contraction. The stimulating electrodes are then moved a measured distance towards the muscle and the experiment repeated. The difference in time between stimulus and response in the two traces

215

can be calculated from the drum speed (or a time trace can be recorded on the drum with a tuning fork or time marker—see kymograph notes).

Procedure

1. Set up the kymograph apparatus to record a muscle twitch.
2. Connect a stimulus marker to the circuit from the induction coil or neon stimulator.
3. Place the stimulating electrodes as close as possible to the end of the nerve. Record a twitch.
4. Move the stimulating electrodes 2 cm closer to the muscle and record another twitch.
5. Measure the difference in reaction time (application of stimulus to the beginning of response). This difference is equal to the time taken by the nerve impulse to travel 2 cm. Calculate the rate of conduction in cm sec.

Requirements

Apparatus

Kymograph
Smoking equipment
Stimulator (induction coil or neon stimulator)
Time marker or 100 cycle tuning fork
Ruler marked in millimetres

Reagents

Frog Ringer
Kymograph varnish

Biological

Pithed frog

An investigation of the excitability of smooth muscle

The smooth muscle of the vertebrate gut is innervated by fibres of the inhibitory, sympathetic nervous system and fibres of the excitatory parasympathetic nervous system. Control of the gut muscle is achieved by the antagonism of these two components of the autonomic nervous system. Nerve endings release noradrenaline (sympathetic fibres), or acetylcholine (parasympathetic fibres). The effects of the autonomic nervous system can be simulated by adding adrenaline or acetylcholine to the Ringer solution which bathes the tissues.

Procedure

1. Set up the kymograph to record at a very slow speed.
2. Kill the rat or mouse with a sharp blow to dislocate the neck. This must be done by the teacher in the preparation room.

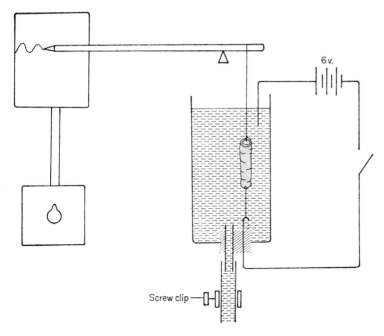

3. Remove the intestine and cut it in 2 and 3 cm lengths. Rinse out the pieces in mammal Ringer, warmed to 37°C.
4. Sew a thread through each end of a piece of intestine, leaving the lumen open.
5. Place the preparation in the perfusion chamber shown in the diagram.
6. If the muscle is contracting rhythmically, record its action and then stimulate it electrically.
7. Drain Ringer from the chamber and run in a Ringer solution containing adrenaline. Record and stimulate electrically if necessary.
8. Drain off Ringer containing adrenaline and fill the chamber with plain Ringer.
9. When normal action has been resumed, replace the Ringer with a Ringer solution containing acetylcholine. Stimulate electrically if necessary.

Note. The length of gut employed is critical. If it is too long, 2 or 4 cycles of peristaltic contraction may be present and cancel each other out so that little or no movement of the recording lever occurs.

Alternative experiment

The above experiment can be carried out using the whole bladder of a frog, tied with cotton at each end or with a portion of frog intestine. If frog material is used, frog Ringer at room temperature is used in the perfusion chamber and in making up the solutions of adrenaline and acetylcholine.

Note. An enzyme, cholinesterase is often present in the tissue and

217

may break down the acetylcholine before it has affected the muscle action. The action of cholinesterase may be inhibited by adding physostigmine sulphate (eserine) to the perfusion medium (1 part in 10^{-4}).

Requirements

Apparatus

Kymograph
Smoking equipment
Perfusion chamber
Isotonic lever
2 or 3 2 volt accumulators
Copper wire
Morse key
Stimulus marker (May be incorporated in the circuit if available)

Reagents

Ringer solution: mammal or frog as appropriate
Acetylcholine solution ($1:10^{-6}$) made up in frog or mammal Ringer
Adrenaline solution ($1:10^{-7}$) made up in frog or mammal Ringer
Kymograph varnish

Biological

Rat or mouse, killed immediately before use
or a frog, pithed immediately before use.

Additional experiments

If satisfactory results are obtained before the end of the practical, try the effect of

1. Single, short shocks.
2. When the muscle is contracting rhythmically try the effect of electrical stimulation at different phases of the cycle of contraction and relaxation.
3. A single shock of 60 seconds duration.
4. A succession of rapid shocks.

An investigation of the neuromuscular system of the locust

A preparation of the metathoracic leg of the locust forms excellent material for the investigation of neuromuscular physiology as the dissection is probably easier than that of the frog gastrocnemius muscle and living material is readily available.

The muscles of arthropods are innervated by 'slow' axons, which produce a contraction if the frequency of stimulation is high, and 'fast' axons which cause a muscle contraction as the result of a single stimulus. Repeated stimulation of the 'fast' axon at a high frequency will bring about a sustained contraction (tetanus). The 'fast' axons probably control the more rapid locomotory movements and the 'slow' fibres probably control slow movements and muscle tonus.

Procedure

1. Etherise a locust.
2. Kill the locust by decapitation. This should be done by the teacher in the preparation room.

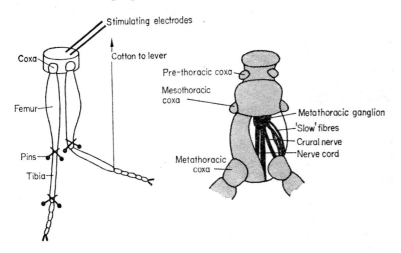

3. With sharp scissors cut off the abdomen, wings and first two pairs of legs.
4. Pull the gut out of the thorax.
5. Pin the preparation, dorsal side down, on the edge of a cork or balsa wood block.
6. Use two pairs of crossed pins to immobilise one metathoracic leg and another pair of pins to immobilise the femur of the second metathoracic leg.
7. Remove a portion of the ventral cuticle of the metathoracic segment to expose the metathoracic ganglion and the nerves which run from it into the coxa of the leg.

Examination of the action of the 'fast' fibres

8. Connect the free tibia to a light lever with a piece of cotton.
9. Place electrodes in contact with the crural nerve and connect to neon stimulator or induction coil.
10. Adjust the stimulator to give the weakest possible stimulus at 1 second intervals.
11. Slowly increase the size of the stimulus until a maximal twitch is obtained.
12. With the stimulator set to give a just maximal twitch, give the preparation short bursts of stimuli (15–20) starting with 1 second intervals and reducing the interval between stimuli gradually until a smooth tetanic contraction occurs.

219

Examination of the action of 'slow' fibres

13. Cut the crural nerve close to the ganglion and the coxa. Remove the severed portion.
14. Place the electrodes in contact with the smaller nerves which lie anteriorly to the crural nerve.
15. What is the effect of a single, strong, stimulus?
16. Give short bursts of 10–20 shocks, starting with a frequency of one shock per second and increasing to 15 per second. What is the effect of increasing the rate of stimulation from 10 per second to 200 per second, moving up in steps of 10?

Requirements

Apparatus

Kymograph
Smoking equipment
Neon stimulator or induction coil
Dissection instruments
Pins
Etherising bottle
Cork block or balsa wood block
Silver or platinum electrodes
Stimulus marker (optional)
Isotonic lever and stand
Make and break key (Morse key)
Two 2 volt accumulators if induction coil is used
Copper wire to complete the circuit
Varnish tray (Dipping tray from Shandon Chromatography kit is ideal)

Reagents

Locust Ringer
Ether

Biological

Locust

The cathode ray oscilloscope

The oscilloscope is an instrument which is used to detect changes in potential which may be both small and rapid. In this respect it is the ideal instrument for following the electrical events which are associated with the passage of a nerve impulse. There are many instruments available which vary considerably in sophistication. For biological work the instrument of choice would be a high quality, twin beam oscilloscope with a built-in preamplifier. Such an instrument is not likely to be found in schools, so the experiments included in this book are designed to be carried out with the normal single beam oscilloscope, which should be found in most school physics departments, and a home made preamplifier. The preamplifier can be built from the circuit given at a cost

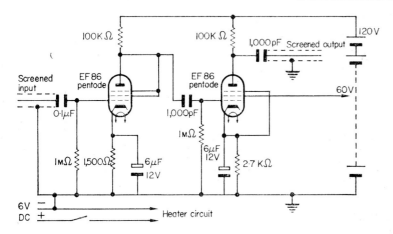

of approximately £2. (The components are all available from Radiospares Ltd, P.O. Box 2BH, 4–8 Maple Street, London, W.1.)

This two-stage amplifier will give a gain of 1,500–2,000 times, depending on the condition of the components of the circuit. For some biological work much higher gain is required, e.g. in recording pace-maker impulses and brain impulses. When the gain of the preamplifier is very high, a great deal of 'noise' is picked up from biological material and the preamplifier needs sophisticated circuitry to separate the noise from the required signal.

Electrodes

In biological work metallic electrodes, usually platinum or silver, or glass, microcapillary electrodes filled with an electrolyte are used. The latter type is used for intracellular measurements on living material. For the experiments given in this book, metallic electrodes are suitable. The potential of a silver or platinum electrode may vary with the halides present in biological tissue, furthermore, if a large current is passed electrolysis and polarisation of the electrodes may occur. For this reason, metallic electrodes are plated with their halides before use in critical work.

Preparation of platinum/platinum chloride electrodes

Clean a short length of platinum wire by immersion in concentrated sulphuric acid. Solder a 2 cm length of platinum wire to a convenient length of copper wire (preferably direct to the core of a coaxial cable). Clean the platinum tip again with concentrated sulphuric acid, rinse in distilled water and immerse in a solution of platinum chloride. Connect the cable to the negative terminal of a 2 volt accumulator, then connect the positive terminal to a second platinum electrode immersed in the solution, completing the circuit with a reversing switch. Pass a current for 15 seconds and then reverse it. Repeat six times, ending with the required electrode positive. The electrode should be washed and stored in distilled water or Ringer.

221

Preparation of silver/silver chloride electrodes

Silver wire (obtainable from jewellers) is degreased in petroleum ether. The cleaned wire is then placed in a N/10 solution of hydrochloric acid and connected to the negative terminal of a 2 volt accumulator via a reversing switch. A second piece of wire is connected to the positive terminal via the reversing switch. The current is passed for 30 seconds and then reversed. The process is repeated three times, ending with the chosen electrode positive. Silver chloride is decomposed by strong light and the plating should be carried out in the dark. Plated electrodes should be stored in distilled water or Ringer in the dark.

Screening

Unless precautions are taken to prevent the pick-up of stray signals such as 50 cycle hum and television frequencies, the 'noise' will obscure the biological event being monitored. The most satisfactory method of screening is to enclose the animal and electrodes in a fine mesh wire-netting 'box'. Cables to the preamplifier and oscilloscope will, of course, be coaxial.

Setting up the oscilloscope

1. Switch on the instrument.
2. When the spot appears, adjust the **focus** and **brightness** controls. **It is very important that a bright, stationary spot is not left sharply focused as it will damage the screen.**
3. The spot when out of focus should not appear as a vertical or horizontal line. If it does the **astigma** control should be adjusted until the effect disappears.
4. Use the **X-shift** and **Y-shift** controls to position the trace.
5. Switch on **time base** control to give a sweep of suitable speed.
6. Connect electrodes through the preamplifier to the oscilloscope.
7. If necessary adjust the **stability** control to obtain a steady trace.

An oscillographic investigation of the sciatic nerve of a frog

This experiment may be carried out on the same preparation as the kymograph experiment on page 211.

Preparation

Prepare a gastrocnemius and sciatic nerve dissection as described on pages 211–212. Keep the preparation moist with frog Ringer

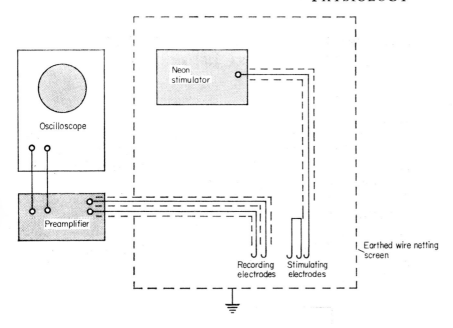

Procedure

1. Place the preparation on a cork board.
2. Lay the nerve across the electrodes from the neon stimulator or induction coil.
3. Place the recording electrodes as far as possible from the stimulating electrodes and connect to the oscilloscope through the preamplifier.
4. Enclose the preparation in an earthed cage of wire netting.
5. Stimulate the preparation and adjust the oscilloscope gain and time base to obtain a good trace.

Requirements

Apparatus

Oscilloscope
Preamplifier
Wire-netting cage
Electrodes
Cork board
Pins
Dissecting instruments
Neon stimulator or induction coil

Reagents

Frog Ringer

Biological

Pithed frog

An investigation of sensory nerve impulses in the cockroach

Procedure

1. Etherise the cockroach.
   Kill a cockroach by decapitation with a sharp scalpel. This should be done by the teacher in the preparation room.
2. Carefully remove the dorsal exoskeleton of the abdomen without damaging the anal cerci.
3. Carefully remove the contents of the abdomen to reveal the ventral nerve cord.
4. Place a pair of silver electrodes in contact with the nerve cord and connect to the oscilloscope through the preamplifier.
5. Screen the preparation with an earthed, wire-netting cage.
6. Touch the anal cerci with a seeker.

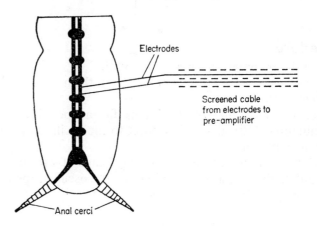

7. Adjust the gain and time base controls of the oscilloscope so that the action potentials produce spikes of 1·5–2 cm height.
8. Bring various tuning forks close to the anal cerci. Are you able to deduce the sound range to which the animal is sensitive?
9. Is there any difference in either the frequency of the action potentials or the number of spikes in a train with different sound stimuli?

Requirements

Apparatus

Oscilloscope
Preamplifier
Electrodes
Wire netting
Dissection dish
Dissection instruments
Pins

Reagents

Insect Ringer

Biological

Cockroaches

ANIMAL BEHAVIOUR
and
PLANT HORMONES

Animal Behaviour

As many theoretical textbooks leave much to be desired in their treatment of animal behaviour, we have decided to include a very brief theoretical account, in general terms, so that the most important behavioural terms which we use later may be seen in context. By so doing we hope that the succeeding experiments will make a coherent introduction to both animal behaviour and some of the methods of studying behaviour patterns.

The value of most experimental work on behaviour depends upon the quality of the observation, often precise measurement is not possible. The experimenter in this field may need to resort to statistical methods to check the significance of his observations.

The behaviour of an organism may be very complex, involving the use of many types of receptor and effector organs, co-ordinated by a more or less complex nervous system and/or the use of hormones. Even if we fully understood the physiology of all the tissues involved, we should still be some way from understanding behavioural patterns. The sensory system, nervous system and the hormones are merely the biological machinery which makes the behaviour patterns of animals possible.

In studying the behaviour patterns of an animal, we should always be conscious of the need to understand how a particular piece of behaviour is **adaptive** to the ends of both the individual and the species.

At one time investigations of behaviour attempted to break down complex patterns into simple reflex arcs in which each stage is triggered, or **released**, by a specific stimulus or **release mechanism (sign stimulus)**. Although much progress has been made in analysing instinctive behaviour and recognising the sign stimuli, it would now seem to be a gross oversimplification to seek such an explanation for all behaviour.

There appears to be a phylogenic development of behaviour patterns, with **stereotyped** behaviour patterns only, in the lower phyla. **Learning** appears in the triploblastic animals and **reasoning** occurs in the mammals. The higher animals tend to show a marked reduction in stereotyped behaviour. Presumably when new behaviour mechanisms are evolved, they exist side by side with the old mechanism for a time and only slowly supersede them.

To make an elementary study of behaviour it is convenient to define and classify the types of behaviour which we observe:

1. Stereotyped

(a) Kineses

The activity rate is directly or inversely proportional to the strength of the stimulus. Animals showing kineses tend to collect in regions

227

	Stereotyped			Not Stereotyped	
	Taxes	Reflex	Instinctive	Learning	Reasoning
Protozoa					
Coelenterates					
Annelids					
Insects					
Anamniotes					
Birds					
Lower mammals					
Lower primates					
Man					

Diagram to indicate a phylogenetic change in the relative importance of different behaviour mechanisms

where they are least active, i.e. regions where the intensity of the stimulus is high or low.

   (i) Klinokinesis: The rate of change of direction of the movement is related to the strength of the stimulus. When in an area of strong stimulation the animals move in a more or less straight line. When the animal reaches an area of low stimulation it turns frequently and tends to remain in the area.

  (ii) Orthokinesis: The speed and/or frequency of movement varies with the strength of the stimulus. Animals tend to congregate in areas where the stimulus strength is low and hence where the rate of movement is least.

 (iii) Akinesis: Animals moving about will stop at some new stimulus, e.g. worker bees moving about in a hive will stop if a high frequency sound is played to them.

(b) Taxes

The movement is not random. The body is orientated either towards (**positive taxis**), or away from (**negative taxis**), the source of stimulation.

   (i) Tropotaxis. Orientation is achieved by the use of bilaterally symmetrical sense organs, the animal turning to one side until both organs are stimulated equally. It involves simultaneous comparison of stimulus strength.

  (ii) Klinotaxis. This involves successive comparisons of stimulus strength, achieved by moving the body, or the sensory organ, from side to side.

 (iii) Telotaxis. Attainment of orientation is direct, the animal 'fixes' the source of stimulation on a part of the sense organ and moves so that the same part of the organ is continuously stimulated throughout the response. So far, only telotaxis towards light has been observed.

(c) Reflexes

These are sometimes difficult to distinguish from taxes but as a general rule taxes are orientation movements involving the whole animal, whereas reflexes usually involve only a part of the body.

- (i) **Tonic reflexes:** Relatively slow adjustments, the effects of which persist for some time, e.g. the reflex changes which maintain posture and muscle tone.
- (ii) **Phasic reflexes:** These are rapid, short-lived adjustments such as cessation of flight in the locust if the legs make contact with a solid object.

(d) Instinctive behaviour

Complex patterns of behaviour which are inherited and typical of the species. Such patterns often depend upon some predisposing internal factor, such as the presence of a sex hormone, before some external, environmental, sign stimulus can trigger off, or release, the complete pattern of behaviour. Sometimes the internal factor may be so strong that the behaviour pattern is released in the absence of a sign stimulus—this is known as a **vaccuum activity.**

2. Learning

This type of behaviour requires modification of an animal's reflexes as the result of experience. There would appear to be a number of different types of learning which differ in complexity.

(a) Imprinting

This occurs mainly in birds and consists of the young bird recognising the first large, moving object which it sees as its 'parent'. Once a 'parent' has been adopted the young bird will follow it in preference to an adult of its own species.

(b) Habituation

On repeated exposure to a meaningless stimulus, the response of an animal gradually reduces and may completely disappear. Habituation enables responses which have no significance in the life of the animal to be dropped. Most forms of learning are the reverse of this process and consist of strengthening responses which are significant.

(c) Conditioned reflexes

In the classical conditioned-reflex, an irrelevant stimulus is associated with the response by repetition until the irrelevant stimulus alone will evoke the response.

229

(d) Instrumental learning

Many forms of acquired behaviour are more complex than the classical conditioned reflex. The animal at first tries out a range of responses which are within its ability. Some of these responses are more appropriate than others and the animal increasingly selects these if the situation is repeated. By selecting responses the animal has some control over its situation and is instrumental to learning. In instrumental learning a reward, such as the satiation of a 'drive', or a failure to achieve the 'goal' act as reinforcements to the response. No irrelevant stimulus is associated with this kind of conditioned behaviour.

An investigation of klinokinesis in 'Planaria'

Procedure

1. Prepare a Petri dish by glueing, or sellotaping, a circle of graph paper to its underside and then filling the dish with water. If three dishes are set up, several sets of results will be obtained and can be averaged.
2. Place one, dark adapted planarian in each dish. Replace lid.
3. Readings should be taken on each dish in turn, recording the movements of the animal and its position every minute. This may be carried out on a piece of graph paper identical to that used in the experimental chamber. Illuminate the first dish with a light which is just bright enough to make readings by. After 5 minutes increase the light intensity to a level which enables the experimenter to see easily. Five minutes later increase the light intensity to a very high level.
4. Return dish 1 to the dark and repeat experiment with dishes 2 and 3.
5. Measure the total angle turned by each animal in each light level, and divide by 5 to give the mean angle of turning per minute.
6. Plot a graph of mean angle of turning against light intensity. (Light intensity can be measured accurately during the experiment by using a photometer or photographic light meter. If a meter is not available, use the same light source throughout the experiment, varying the intensity by altering the distance of the lamp from the animal; light varies inversely as the square of the distance, i.e. a light source at 2 feet covers four times the area which it would cover at 1 foot, and sixteen times the area at 4 feet.)

230

Requirements

Apparatus

3 Petri dishes with lids
Several sheets of graph paper
Light sources or bright light with rheostat
Photometer or light meter if available
Darkroom or walk-in cupboard

Biological

Dark adapted planaria (kept for at least 1 hour in total darkness)

An investigation of orthokinesis in woodlice (Method 1)

Woodlice are particularly sensitive to changes in relative humidity and tend to be more active in a dry atmosphere than in a moist atmosphere. If they are placed in a choice chamber with low humidity on one side and high humidity in the other, they may be expected to move into the more humid chamber. In choice chamber experiments with a limited number of trials and animals, 'perfect' results are unlikely to occur. Statistical examination of the results will help us to decide whether the figures we obtain are sufficiently different to be significant.

Procedure

1. Place a large wad of cotton wool, which has been soaked in water and then squeezed out, at one end of a large glass trough (or a long Perspex box). At the other end of the trough or box place a container of calcium chloride. Support a piece of perforated zinc, cut to fit the trough or box, just above the cotton wool and calcium chloride. The apparatus is then fitted with a

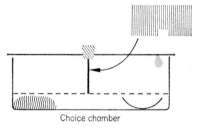

Choice chamber

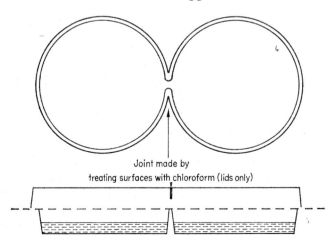

Joint made by
treating surfaces with chloroform (lids only)

231

close-fitting Perspex lid with a central hole closed by means of a rubber bung.

2. Place 10 woodlice, of the same species in the chamber by dropping them through the bung-hole, after the apparatus has been allowed to stand for 10 minutes. (This is essential to obtain a humidity gradient.)
3. After 5 or 10 minutes, count the number of animals in each half of the chamber.
4. Remove the animals and repeat the experiment.

Statistical procedure

If we assume that humidity has no effect on the distribution of woodlice, equal numbers of animals should be found on either side if sufficiently large numbers of animals or trials were used. If smaller numbers are used, it is improbable that equal numbers will be found in each half. We need to check our numbers statistically, using a quantity known as the **Standard error**, which is related to the expected proportion (in this instance $0.5 : 0.5$) and the number of trials made. If the expected proportion is $p : q$, where $p + q = 1$, and we have used n animals in the trial, the Standard error of the proportion is

$$\text{S.E.} = \sqrt{\left(\frac{p : q}{n}\right)}$$

If we apply this formula to our **null hypothesis** after 10 trials of 10 woodlice $p = q = 0.5$ and $n = 10 \times 10 = 100$.

$$\text{S.E.} = \sqrt{\left(\frac{0.5 \times 0.5}{100}\right)} = 0.05$$

If the counts in our experiment were 80 on the humid side and 20 on the dry side, the observed ratio is $0.8 : 0.2$. These figures differ from the expected ratio by 0.3. We are now able to check our results using a standard set of statistical tables known as 't' tables (see appendix).

$$\text{The statistic '}t\text{'} = \frac{\text{Difference from expected ratio}}{\text{Standard error}}$$

$$\text{in this case } t = \frac{0.3}{0.05} = 6$$

We now look in the t tables in the line where $n = 100$ until 6 is found. Six is in fact off the table, which means that the probability of our results being insignificant is less than 0.1%. This means that there is only a one chance in a thousand that our original null hypothesis is correct.

From this we may reasonably assume that the differences in numbers of woodlice on either side of the choice chamber is due to the difference in humidity. It is conventional to accept a probability of 5% as being significant. In other words, we would be right in 19 cases out of 20. If a figure very close to 5% is obtained, it may be desirable to carry out a greater number of experiments to be as accurate as possible.

Requirements

Apparatus

Large glass trough or Perspex choice box
Perforated zinc to fit the above
Cotton wool
Small container for calcium chloride
Perspex cover and rubber bung for trough or choice chamber
t tables

Reagents

Calcium chloride

Biological

10 woodlice of the same species

An investigation of orthokinesis in woodlice (Method 2)

Please read the theory section of the previous experiment. The relative humidity of a closed vessel may be prearranged by placing in it an appropriate mixture of glycerol and water as shown in the table.

% R.H.	40	50	60	70	80	90
% glycerol	85·5	80·0	70·35	64·0	50·15	20·75

Procedure

1. Set up six crystallising dishes, each with a piece of perforated zinc supported above the appropriate solution of glycerol. Cover each dish with glass or Perspex.
2. Place 10 woodlice in three of the chambers.
3. Allow 10 minutes for the animals and the humidity to settle down, then count the number of animals moving in each chamber. Record the temperature.
4. Transfer the animals to the remaining three chambers and allow 10 minutes to settle down. Count the numbers moving in each chamber. Record the temperature—if it has changed since the first three chambers were counted, the results may be invalid. Fluctuations caused by draughts in the laboratory must be avoided.
5. Are you able to observe a relationship between the number of animals moving in each chamber and the relative humidity? If the answer is yes, plot your results as a graph or as a histogram.

233

Requirements

Apparatus

6 crystallising dishes (without lip)
6 glass or Perspex covers
6 pieces of perforated zinc cut to fit the dishes
Some form of support to keep the zinc above the liquid
Thermometer

Reagent

Glycerol diluted as shown in the table on page 233

Biological

At least 30 woodlice of the same species

An investigation of akinesis in centipedes

When centipedes are disturbed, the majority of species run about very rapidly. The activity continues until the animal feels contact on all sides. The activity then ceases abruptly.

Procedure

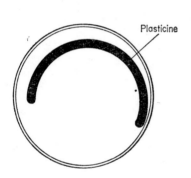

Plasticine

1. Place a piece of rolled plasticine, about $\frac{1}{4}$ of an inch in diameter and $1\frac{1}{2}$ inches long, parallel to the edge of a Petri dish. In the first instance the plasticine should be placed about $\frac{1}{2}$ an inch from the edge of the Petri dish.
2. Introduce a centipede to the Petri dish. If the centipede is able to run between the plasticine and the edge of the dish, the gap should be progressively reduced until the animal is unable to pass through the gap without touching both the plasticine and the dish.

Requirements

Apparatus

Petri dish
Small piece of plasticene

Biological material

Centipede

An investigation of akinesis in the honeybee

When a bee colony possesses a queen, the bees show steady activity when observed in an observation hive. If the queen is removed, activity becomes disorganised and rapid rushing about is observed. This does not occur if mature queen cells are present in the colony, probably because the young queens emit a high pitched 'piping' sound which indicates their presence. If a wet finger is drawn sharply across the glass of the observation hive, all activity will cease during the activity of the squeak.

Procedure

1. Observe akinesis when a wet finger is drawn across the glass of an observation hive.
2. Attach a small vibrator or loudspeaker to the hive.
3. Connect the vibrator or speaker to the output terminals of a signal generator.
4. Start with a note of low frequency and increase the frequency slowly until akinesis is observed. What happens if the frequency is increased further?

Requirements
Apparatus

Observation hive containing bees   Signal generator
Vibrator or loudspeaker   Wire to connect

An investigation of tropotaxis of woodlice in light

Procedure

Examine the behaviour of woodlice in the following lighting conditions:

(i) Weak, diffused light.
(ii) Illuminated from above with a single lamp to give a radial intensity gradient. For this experiment, use 20 woodlice.
(iii) Set up two lamps at right angles to each other. Observe the reactions of the woodlice when (*a*) both lamps are switched on, and (*b*) when the lamps are switched on alternately.
(iv) Cover one eye of a woodlouse by painting it with a quick drying black paint (model aircraft, cellulose paint is suitable). Place the animal in a single light beam and record its movements.

Requirements
Apparatus

2 bench lamps
Darkroom or walk-in cupboard
Fine paint brush

Reagents

Black paint (quick drying)
Xylene or ethyl acetate to clean brush

Biological

Woodlice

235

An investigation of klinotaxis in blowfly larvae

This experiment must be carried out in a darkroom, using dark-adapted animals.

Procedure

A few days before they pupate, blowfly larvae become strongly photonegative. As the photoreceptors are on the maxillary lobes, they are screened from the light source by the larva. To overcome this, the animal moves its head from side to side, so that successive comparisons of light intensity on each side of the body can be made.

Experiment 1

1. Place 20–30 dark adapted larvae in a heap near a 60 watt light bulb.
2. Count the number of larvae which move away from, and towards, the light.

Experiment 2

1. Lay a piece of smoked kymograph paper on the bench.
2. Set up two bench lamps, each with the same power bulbs, so that their beams are at right angles to each other. Make sure that there are no other light sources.
3. Place a larva on the paper, 50 cm from the lamps. Switch on one lamp.
4. Switch off the first lamp and switch on the second.
5. Switch on both lamps.
6. Repeat using different larvae.
7. Varnish the trace, after adding necessary information by writing with a seeker point.

Requirements

Apparatus

Kymograph paper
Kymograph drum
Smoking equipment
2 bench lamps with 40 or 60 watt bulbs

Biological

Dark adapted blowfly larvae. (The larvae should have been kept in a jar of dry sawdust, in complete darkness for several hours before use)

An investigation of telotaxis in the earwig ('Forficula')

Procedure

Examine the behaviour of earwigs in the following lighting conditions:

   (i) Weak, diffused light.
   (ii) Illuminated from above to give a radial intensity gradient. Use 20 earwigs if available. If trouble is experienced with escapes, use fewer animals, or carry out the experiment in a deep-sided tray with vaselined edges.
   (iii) Set up two lamps at right angles to each other. Observe the reactions of earwigs when (*a*) both lamps are switched on, and (*b*) when the lamps are switched on alternately.
   (iv) Cover one eye of an earwig by painting it with a quick drying black paint (model aircraft cellulose). If necessary, the earwig can be lightly etherised for painting but must be given at least ten minutes after resuming activity, to return to normal behaviour. Alternatively the earwig may be placed in a refrigerator for 10 minutes before painting.

In what respects did the behaviour of your earwig differ from the behaviour of the woodlice used in the previous experiment?

Requirements

Apparatus

2 bench lamps
Darkroom or walk-in cupboard
Fine paint brush
Etherising bottle, as used in *Drosophila* work, or refrigerator
Tray and Vaseline (optional)

Reagents

Black paint
Xylene or ethyl acetate to clean the brush
Ether

Biological

Earwigs

An investigation of the light responses of adult blowflies

Procedure

1. Investigate the behaviour of a dark adapted, de-winged blow-fly when it is placed in a single light beam. Repeat several times using different flies.

2. Investigate the responses of a dark adapted, de-winged blow-fly when it is placed in the beams of two lights, at right angles to each other. The lamps should be switched on and off alternately.
3. Paint the eye on one side of a de-winged, dark adapted blowfly, with black paint. Examine its responses in a single light beam.
4. Compare the behaviour of a de-winged blowfly, with that of a de-winged blowfly which has had one eye covered with black paint when both are placed underneath a single lamp.

Results

When you analyse the results of your experiments, do you conclude that adult blowflies respond to light tropotactically or telotactically?

Requirements

Apparatus

2 bench lamps
Fine paint brush
Etherising bottle or refrigerator to immobilise flies for painting

Reagents

Paint
Xylene or ethyl acetate to clean brush
Ether

Biological material

Adult blowflies which have been de-winged (immobilise by chilling in refrigerator). The wings are cut off close to the thorax. The prepared flies should be dark adapted by keeping them in the dark for 1 hour before use.

Investigation of reflex actions

Although reflexes play an important part in the behaviour of many animals they are very difficult to study scientifically under school conditions. In vertebrates, the majority of adaptive reflexes are subject to modification by the higher centres of the central nervous system. To be certain that one is observing a spinal reflex it is usually necessary to sever the spinal cord below the brain. Reflex actions are adaptive and result in either relatively slow changes affecting posture, muscle tone, pigment distribution (**tonic reflexes**) or rapid movement of a part of the body in response to a stimulus (**phasic reflexes**). Often reflexes are grouped in chains.

Because of the experimental problems involved and the difficulty of carrying out valid experiments without infringing the vivisection Acts, the following experiments are designed to demonstrate, rather than investigate, reflex actions.

Experiment 1. The human knee-jerk reflex

This is a rather dramatic tonic reflex. The state of contraction of a muscle is monitored by sensory spindles which inform the spinal cord of changes in the tension of the muscle. This is a mono-synaptic reflex arc, i.e. the sensory neuron connects directly with the motor neuron without the interpolation of an internuncial neuron. If the tension of the muscle is changed by stretching it, an impulse from the spindle passes to the spinal cord and a motor impulse is initiated which brings about contraction of the muscle. One way of applying a rapid stretch stimulus is to give the patellar tendon a sharp tap. This causes a small, rapid extension of the quadriceps muscle of the thigh, followed by a sharp contraction of the quadriceps muscle.

1. Seat the subject with the legs crossed, so that one leg swings free below the knee with its quadriceps muscle under light, constant tension.
2. Strike the patellar tendon sharply using a patellar hammer, spine of a book, or the edge of the hand.

Experiment 2. Colour changes in fish

Some species of fish show an adaptive colour change to match their environment. This is a slow tonic reflex involving a visual stimulus. The response is brought about by a change in the distribution of pigment in the chromatophores.

1. Prepare three small, glass aquaria as follows:
   (a) Surround three sides and the bottom with black paper.
   (b) Surround three sides and the bottom with white paper or aluminium foil.
   (c) Surround this tank with paper which has been painted to give alternate, vertical stripes of black and white (3 stripes to the inch approximately).
2. Fill the tanks with water by siphoning from the larger aquarium containing the stock of fish. As the fish used in this experiment are tropical fish it is essential that they are not subjected to sudden changes in temperature.
3. Introduce one angel-fish (available from local aquarist shops) to each tank. **Always use an aquarium net when transferring fish as hands are more likely to damage the protective surface mucilage and lead to fungal attack.**
4. Observe any colour changes carefully. How long is it before noticeable changes occur? How long does complete adaptation take?
5. Return fish to their heated aquarium gently **as soon as possible after the experiment.**

Note. Angel fish have been selected for this experiment as they show a striking adaptation in the circumstances described. They sometimes show rapid colour changes when handled or frightened —presumably as the result of adrenaline action. If angel-fish are not available in the school aquaria, the experiment may be tried with any other species, using two tanks, one with a black background, the other white. If you are uncertain whether or not adaptation has occurred, pour one fish into the other tank or exchange the paper backgrounds so that the two fish can be compared for slight colour differences.

Experiment 3. Flight and landing reflexes in locusts

The flying action of locusts is elicited if a suspended animal is stimulated by blowing on it. As long as a locust is in an airstream, with its feet out of contact with the ground, it will fly continuously. If a solid is placed in contact with the feet of the locust it will cease to fly.

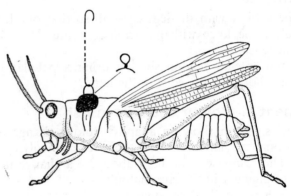

1. Attach a small ring of light wire to the thorax of an adult locust, using a spot of wax.
2. Suspend the locust on a wire hook from a retort stand.
3. Blow gently on the head of the locust. Observe reaction on (*a*) application and (*b*) cessation of the stimulus.
4. Bring a fan into position so that the locust receives a constant airstream.
5. When the locust has been flying steadily in the airstream for some time, place a piece of cotton wool or paper in contact with its feet.

The landing response on contact is stronger than the flying response to an airstream. What is the adaptive significance of this?

Instinctive behaviour patterns

The concept of instinctive behaviour has been somewhat obscured by the way in which many behaviour patterns, even in humans, have been explained in terms of instinct, often without any investigation of the underlying mechanisms. Possibly one of the factors which has led to overuse of the instinct concept is the difficulty of producing criteria to define instinct. Generally speaking, instincts are characteristic of the species, adaptive and unlearned. In applying these criteria the difficulties begin. All forms of behaviour are adaptive and in the higher animals it is often very difficult to prevent any chance of learning by experience and observation.

Examples of instinctive behaviour patterns which may be readily observed in the school laboratory

1. Keep young, mature males and females of the red swordtail apart for several days. Then introduce a male into a tank containing the female. An instinctive mating ritual will be observed, the male swimming rapidly backwards and forwards in front of the female. After a while the female adopts a submissive posture, the normally erect dorsal fin droops and the female remains more or less stationary. The male then darts backwards towards the female, inserting a gonopodium of modified fins into the cloaca of the female to effect mating. (Young red swordtails are usually available from aquarists shops at about 5/- per pair. The sexes are easily distinguished as the male is smaller and possesses a sword-like extension of the lower part of the tail fin.)

2. Male Siamese fighting fish are intolerant of each other and will inflict very heavy injuries on an intruder. The fighting behaviour can be released if a mirror is gently lowered into an aquarium containing a male fish. The fish will vigorously attack its own reflection. If a large aquarium 36 in × 12 in × 15 in is available it is possible to introduce two males to it simultaneously, provided that they are separated by a glass partition. After several weeks the partition may be removed and the fish will probably only fight if one strays into the territory of the other. The visitor usually retreats rapidly to his own territory. It is not wise to leave the males together without a partition, unless they are under observation.

3. If three-spined sticklebacks are kept in the aquarium during the spring and summer terms, the complex mating behaviour, described and analysed by Tinbergen, may be observed.

Egg-laying in locusts: an instinctive pattern

Under normal conditions, locusts lay their eggs in 'pods' at a depth of about 4 inches in moist sand. The behaviour is instinctive and consists of a chain of reflexes. If the necessary trigger mechanism is missing at some stage in the chain, the 'drive' may not be sufficient to take the instinctive pattern to its conclusion. For example, if only dry sand is available the female will insert her abdomen but not lay eggs. If no suitable oviposition site is available, the female may show 'vacuum activity' and scatter eggs over the surface of the cage.

Experiment 1. Observation of oviposition

Make a container from two pieces of glass about 6 in high, by placing a strip of wood $\frac{3}{8}$ in wide at either end. Fill the space to a depth of 4 in with moist sand. Place 2 or 3 female locusts which have been deprived of suitable oviposition sites for 5 days, in the cell and cover to prevent escape. Observe carefully. How do you think the $1\frac{1}{2}$ inch abdomen excavates to a depth of 4 inches?

Experiment 2. Investigation of moisture preference

Set up a standard locust cage (see appendix) with a number of egg tubes containing sand and water in the following proportions:

Tube	Sand	Water
1	100 parts	0 parts by volume
2	100 ,,	5 ,, ,, ,,
3	100 ,,	10 ,, ,, ,,
4	100 ,,	15 ,, ,, ,,
5	100 ,,	20 ,, ,, ,,
6	100 ,,	25 ,, ,, ,,

Place 6 to 8 females of egg-laying age in the cage. Remove tubes after 24 hours and count the number of egg pods in each. Can you express your results as a histogram?

Experiment 3. Investigation of moisture preference (Method 2)

Divide three small pneumatic troughs or 1 litre beakers, by a vertical partition 4 inches deep. Place sand with different moisture content on each side of the partition until it is level with the partition.

Make up the sand/water mixtures as in the above table and set out choices as follows:

Chamber 1 Dry/5 parts water	Chamber 2 5 parts/10 Parts water	Chamber 3 10 parts/15 parts water

Place 6 to 8 egg-laying females in each chamber. Count the egg pods after 24 hours. Use the t test to check the significance of any differences which you observe. The test can, of course only be used to test the significance of pairs of results from the same choice chamber.

Experiment 4. To show that the complete sequence of oviposition behaviour may be broken if conditions are unsuitable

Prepare egg containers by filling them with oven dried sand to a level $\frac{1}{4}$ inch from the top. Fill the last $\frac{1}{4}$ inch with sand which has been stained with methylene blue and then dried. Place the egg containers in a cage which has been deprived of oviposition sites for 24 hours. If the blue sand is carefully removed from the surface, trial oviposition holes will be apparent on account of the blue sand which falls in to the holes when they are abandoned. Note that the oviposition 'drive' has not been satisfied by the laying of eggs.

Experiment 5. Gregarious oviposition in 'Schistocerca gregaria'

Place 4 inches of sand (100 parts sand/15 parts water by volume), in a large pneumatic trough. Tether a live female locust in one segment of the trough, place a paper decoy in the second segment, and leave the third segment vacant. Introduce 3 laying females to the centre of the chamber and leave undisturbed for 24 hours. In which segment are the majority of egg pods deposited?

Is the gregarious behaviour dependent upon vision or a chemical response? Set up 2 identical troughs but place one in total darkness for 24 hours, the other is kept in light. Is there any difference in oviposition in the 2 vessels?

Requirements (For the series of 5 experiments)

Apparatus

2 pieces of glass approximately 6 in × 12 in
2 pieces of wood $\frac{3}{8}$ in × 1 in × 6 in
Sellotape
Locust cage(s)
Egg tubes
Measuring cylinders
Small troughs or 1 litre beakers
Cardboard to make divisions
Large pneumatic trough or polythene wash-bowl at least 5 in deep
Bench lamp
Cotton and stakes

Reagents

Dried sand
Dried sand which has been stained with methylene blue prior to drying

Biological material

Large number of mature egg-laying locusts

An investigation of learning in the rat

Mammals are capable of learning to find food in a maze with only a relatively small amount of training. Seeking food is a primary 'drive' and the effort devoted to the search depends on the degree of motivation. If the animal has recently eaten the 'drive' is said to be satiated and little effort may be spent searching for food. For learning experiments which depend upon the drive to find food, the animals should be motivated by starving them for 24 hours. Measurement of learning is difficult and this experiment has been selected because it provides two methods of assessing learning, namely, time to complete the course, and the number of errors made in each trial. In a two period practical there should be no difficulty in obtaining sufficient results to plot a learning curve. Do not be surprised if a learning curve bears little resemblance to the graphs which you have plotted previously—learning is a complex process and the graph often shows steps, or 'plateaux', e.g. rats may steadily improve their time over a maze until all errors are eliminated, then performance remains static for a while and then improves suddenly as the rat learns to position itself to take the bends of the maze at maximum speed. Plateaux may also be associated with learning to make the right choice at particular junctions in the maze.

Rats which have been used for a maze experiment, normally relearn a given maze with far fewer trials when used for a second experiment—there is some retention. Maze trained rats show a superior performance to untrained rats even if they are compared in a new maze of different construction.

Procedure

1. Place a 'motivated' rat at the start of the maze. Food is placed at the finish.
2. Count the number of errors (wrong turns) made by the rat and time it with a stopwatch.
3. When the rat reaches the food, allow it to eat a very small amount before returning it to the start.

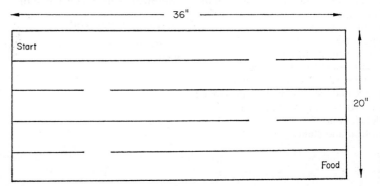

244

4. Repeat the experiment as often as possible, recording time and number of errors as well as any observations of the animal's behaviour.
5. If possible repeat the experiment with the same animal under the same conditions, one week later.
6. Plot two graphs
   (*a*) Number of trial against time
   (*b*) Number of trial against number of errors.

Notes

1. This experiment is best carried out on an animal which is used to being handled.
2. Handle the animal gently, particularly when removing it from the food—if the animal is frightened or frustrated too much, it will show 'displacement activity'. In the rat, displacement activity consists of sitting down and washing the face.
3. During a trial the animal must not be distracted, the experiment is best done with only two experimenters present in the room. Avoid rapid or unnecessary movements.

Requirements

Apparatus

Maze constructed as shown in the diagram
The sides of the maze are 3–4 inches deep and the whole maze is covered with a removable pane of glass
Stop watch
Tally counter (optional)

Reagents

Pelleted food

Biological

Motivated rats, i.e. animals which have not had food in their cages overnight, should be used just prior to normal feeding time. Several rats should be prepared.
N.B. The deliberate starvation of animals for long periods would constitute an 'Experiment' as defined by the Cruelty to Animals Act (1876), and must be avoided.

An investigation of the acquisition of a complex skill in man

In man the acquisition of a simple, single skill involving a single conditioned reflex may be very rapid. When a complex skill such as typing is learned, involving the building up of many new reflexes, improvement may be gradual with a number of well-defined steps of 'plateaux'. Gradual improvement occurs as the subject learns the position of the keys; plateaux appear as the subject acquires the ability to type groups of letters which occur together as a single unit, e.g. 'it is', 'the', 'and'.

Procedure

1. Select a subject with no typing experience.
2. Seat the subject in front of a typewriter—preferably in a room with only the experimenter present to avoid distraction.
3. Ask the subject to type 'Hey diddle diddle, the cow jumped over the moon'.
4. Repeat 100 times, recording the time taken for each trial. If possible note when any groups of letters are fired off more rapidly for the first time.
5. Plot trial number against time on a graph.
   If you have not obtained a smooth learning curve can you explain the irregularities? Is the graph a picture of one or more phenomena?

If two typewriters are available another subject, also without typing experience may also try the experiment, but should alternate 5 minutes practice with 2 minutes rest. Is continuous effort or short efforts with rest periods most effective in a learning situation? N.B. A single experiment of this type may produce results which are difficult or even impossible to draw valid conclusions from. Ideally a large number of subjects should be assessed in each group, some of the largest variables such as intelligence, aptitude and concentration may be ironed out to some extent by redesigning the experiment so that each subject types the same sentence for half of the time available, one resting every five minutes, the other working continuously. The subjects then change learning methods and are given a new sentence e.g. 'It's a long, long way to Tipperary', for the second half of the experiment.

Requirements

Apparatus

1 or 2 typewriters
Stopwatches

An investigation of the interaction of incompatible behaviour mechanisms

The overall behaviour of an animal is governed by the relative strengths of its **'drives'**, e.g. in a well fed, mature animal the **'sex drive'** may be the predominant factor in the behaviour which the animal exhibits. If the same animal is deprived of food for any length of time its behaviour may be predominantly influenced by the **'hunger drive'.** The relative strengths of two incompatible behaviour mechanisms may be compared by placing the animal in a choice situation where the means of satiating the incompatible drives are both present.

Procedure

1. Prepare a box about 2 feet long by 1 foot wide and 4–5 inches deep by placing a 40 watt bulb at one end and food at the other. The box should have a glass top so that the animals can be observed.
2. Introduce 10 well-fed locust hoppers into the box. Note their positions after 5 minutes.
3. Remove the hoppers and replace them with 10 hoppers which have not been fed for 3 hours. Record positions after 5 minutes.
4. Repeat using locust hoppers which have been starved for 6 hours.
5. Repeat with hoppers which have been starved for 12 hours.
6. Repeat with hoppers which have been starved for 24 hours.

Results

Locusts like to bask in areas of high temperature (up to 40°C). Even when very hungry they may show some reluctance to move from a warm to a very cold area for feeding.

Requirements

Apparatus

Arena constructed as described above

Biological materials

10 well fed locust hoppers
10 locust hoppers starved for 3 hours
10 locust hoppers starved for 6 hours
10 locust hoppers starved for 12 hours
10 locust hoppers starved for 24 hours

An investigation to determine whether a left or right bias occurs in an adult locust

When an animal is presented with a choice between two identical alternatives, we would expect that if sufficient trials are considered, equal proportions of each choice would be made. However, in some situations the proportions of choice differ from the expected 50 : 50 ratio by a statistically significant amount.

247

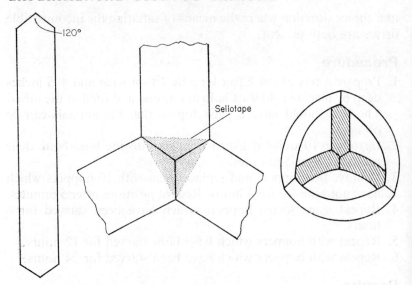

Procedure

1. Make a left/right, simple choice maze as shown in the diagrams, making the joints with triangles of Sellotape placed on both sides of the paper. The paper struts used to construct this apparatus are cut from thick paper or thin Bristol board. The six struts are cut between 2 and 3 inches in length, and $\frac{3}{8}-\frac{1}{2}$ in wide. (The apparatus is known as a spangenglobus.)
2. Attach a 2 mm diameter ring made from fuse wire or cotton to the dorsal surface of the prothoracic segment of an adult locust, using melted wax.
3. Suspend the locust from a retort stand with a piece of stiff wire.
4. Place the spangenglobus in contact with the feet of the locust.
5. As the animal walks (make sure that there is no strong uneven illumination), count the number of right and left hand turns made.
6. Check the significance of your results using the t test.
7. If time permits the experiment may be repeated using the same animal illuminated with a 60 watt bulb on one side. Does this affect the directional choice?

Requirements

Apparatus

Spangenglobus
Retort stand
Fuse wire
Mounted needle
Dissection wax
Bench lamp

Biological material

Adult locust

248

An investigation of the physical changes which occur during emotional stress

When a mammal undergoes stress of some kind a number of changes occur in the body. Many of these changes result from a slight increase in the adrenaline level. Several physical changes can be monitored, changes in blood pressure, breathing rate and the electrical resistance of the skin. In this experiment, an instrument known as the psychogalvanometer is used to monitor changes in the resistance of the skin which are associated with small changes in the amount of perspiration produced. Although we are often able to tell a lie without any visible sign of stress, slight changes in skin resistance still occur.

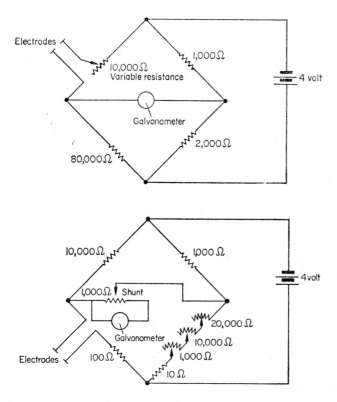

The psychogalvanometer (i.e. Wheatstone bridge connected to galvanometer)

Two circuits for making up the equipment are given, the first is inexpensive and all the materials should be available in a school physics department. The second circuit is a little more expensive in components, but more sophisticated. The galvanometer used may be of the needle type or a mirror galvanometer with a light

249

spot scale, the latter being much more accurate. The electrodes are discs of sheet copper about 2 cm diameter soldered to a copper wire. A disc of chamois leather, soaked in 3 molar potassium chloride solution is placed between the electrode and the skin to ensure good electrical contact.

Procedure

1. Thoroughly wash the subject's hands to remove all dirt and grease.
2. Attach the electrodes, one to the palm, and one to the back of the subject's hand. The electrodes may be held in place by a band of surgical plaster.
3. Make the subject as comfortable as possible with the arm immobilised to prevent false readings caused by muscular action.
4. Allow 5 minutes for the subject's resistance to settle down to its normal level.
5. Adjust the variable resistances to give a galvanometer reading which fluctuates around the zero mark. If circuit 2 is being used the shunt should be adjusted to make the galvanometer as steady as possible with the subject at rest.

Experiment 1

1. Carry out a conversation with the subject, avoiding, as far as possible, any topics or questions which might have emotional overtones. Note the normal range of galvanometer deflection. Ask a series of questions which require a 'yes' or 'no' answer, at 30 second intervals, the subject having been previously asked to answer some truthfully and to try to conceal the truth in some, e.g. ask the subject to conceal the month of his birth when asked 'Were you born in March?'; 'April?' etc.

   Note the galvanometer deflection at each answer. It may be necessary on occasion to adjust the variable resistance to obtain a reading during some answers.

Note. A lie detector of this type cannot be regarded as completely accurate, particularly when it is used by anyone other than a trained, experienced, psychologist. The author has memories of one experiment where a young, female subject deflected the galvanometer spot right off the scale even though no question had been asked. After several repetitions it was observed that the phenomena was related to the movements, in and out of the laboratory, of a handsome lecturer!

Experiment 2

1. Prepare a list of words, e.g.

chair	hammer	girl (boy) friend
cup	car	school
screwdriver	pen	bread

2. Ask the subject to say the first word which comes into his mind when presented with a word from the list. Record the galvanometer deflection and the time taken to answer.

Requirements

Apparatus

Psychogalvanometer made up as in one of the circuit diagrams given
Stopwatch

Reagents

3 Molar solution of potassium chloride (40·6 g per 100 ml)

Examination of insect flight

Insects move their wings with great rapidity during flight so that rate of wing beat and the action of the wing cannot be observed by direct visual means. Butterflies may beat their wings at only 5–10 beats per second. Wing beat rates as high as 1,000 beats per second may occur in very small Diptera. As the wing beat of an insect is a regular affair it can be investigated by stroboscopic means. A stroboscopic lamp consists of a discharge tube to provide flashes of short duration with an electronic circuit which controls the interval between flashes. This interval is varied by means of a potentiometer or a multiposition switch with a number of resistors of different values. If the school does not possess a stroboscopic lamp in the physics department, one can be easily constructed at low cost by the electronics enthusiast. A home-made instrument can be calibrated by using it to observe a vibrator which is receiving impulses from a pulse generator. When the lamp is flashing at the same speed as the wingbeat, the wings appear stationary. (The same effect occurs if the strobing is slower and a 'sample' is taken every second or third cycle of the wingbeat.)

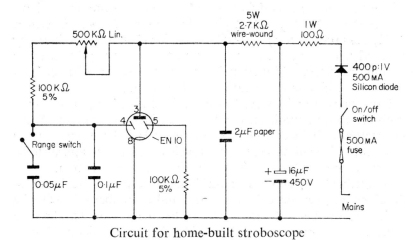

Circuit for home-built stroboscope

251

Procedure

1. Attach a 2 mm diameter ring of fuse wire or thread to the dorsal surface of the prothoracic segment of an adult locust. The ring is most easily attached with molten wax (dip a hot needle into the wax of a dissecting dish).
2. Suspend the locust from a retort stand in a darkroom with a piece of wire.
3. A draught from a small fan or hair drier will stimulate the locust to fly.
4. Switch on the stroboscopic lamp and adjust the rate of discharge until the wings appear stationary. Read the number of cycles from the dial of the instrument or from its calibration curve.
5. Adjust the stroboscope rate until the wings appear to be moving in very slow motion. Do the wings beat independently or together?
   What pattern do the wing tips follow.
6. Repeat with any other insects which are available.

If a stroboscopic lamp is not available, the experiment may be performed with the simple home-made apparatus described in *Looking at Animals Again*, Don R. Arthur; Freeman 1966.

Requirements

Apparatus	**Biological**
Stroboscopic lamp; retort stand; wax; mounted needle 15 amp fuse wire Fan or hair drier	Adult locust

---

Animal hormones

There have been many definitions of the word 'hormone', perhaps the most satisfactory is that given by Jenkin (*Animal Hormones*, 1962, Pergamon Press)—'specific organic substances produced by isolated cells, or by a tissue which may form a gland; they activate or inhibit effects of functional value occurring in other cells or tissues, to which they are carried by the blood'.

Hormones are widespread in the animal kingdom, occurring in both vertebrates and invertebrates in considerable numbers. Some invertebrate phyla and classes appear to lack hormones and in some primitive animals substances such as adrenaline appear to be present as by-products with no functions in co-ordination.

Invertebrate hormones control metamorphosis and sexual development, certain colour changes and a number of metabolic activities. In the vertebrates, the endocrine system has become much more complex, and to a considerable extent, hierarchical.

Practical investigation of hormones is rather difficult as it often involves considerable operative skill and the use of adequate controls. Many investigations of hormone action, even if technically possible in schools, would be undesirable on both ethical and legal grounds. Stages in the investigation of a suspected hormone are:

1. Removal of the suspected endocrine organ and study of the effects of its removal.
2. Implantation of the suspected gland or injections of its extract, into an animal from which the gland has been extirpated.
3. Histological investigation of the gland under varied experimental conditions.
4. Isolation and chemical identification of the hormone.

Whenever operations are carried out on the animal a 'mock' operation has to be carried out on a control.

Jenkin classifies hormones according to their functional role:

1. Kinetic, or control of effectors.
2. Metabolic, or the control of cell biochemistry.
3. Morphogenetic, or the control of growth and differentiation.

The action of a kinetic hormone may be seen in the experiment described elsewhere in this book, when adrenaline is applied to the frog gastrocnemius preparation. Another example, the effect of pituitrin on frog chromatophores is also given below.

The action of metabolic hormones is less easily demonstrated under school conditions but two well known examples may be cited:

(a) The action of insulin.
(b) If the eyestalks of *Carcinus* are removed, or ligatured at the base, the animal loses its osmoregulatory ability.

The action of a morphogenetic hormone in blowfly larvae is detailed below.

Morphogenetic hormones in 'Calliphora'

In the class *Insecta*, development is controlled by several hormones. Juvenile hormone, JH, is released by the corpora cardiaca and the prothoracic glands produce a second hormone, growth and development hormone or GDH as well as the third hormone ecdysone which controls growth and moulting.

If JH and GDH are present at the same time, the insect moults but remains larval. Pupation is initiated when JH ceases to be secreted and GDH alone is present.

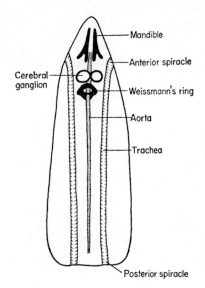

Labels on diagram:
- Mandible
- Anterior spiracle
- Cerebral ganglion
- Weissmann's ring
- Aorta
- Trachea
- Posterior spiracle

In the Diptera the corpora cardiaca, corpora allata and pro-thoracic glands are fused together to form a ring, known as Weissmann's ring, surrounding the aorta immediately posterior to the brain. By ligaturing larvae in different places, the secretions from this ring of endocrine tissue can be studied.

Procedure

1. Select 24 larvae which are just fully grown (post-feeding larvae).
2. Divide into 2 groups of 12 larvae and set 1 group aside for 3 days.
3. The other group is subdivided and treated as follows:
   (*a*) Control—no treatment.
   (*b*) Tie a cotton ligature anterior to Weissmann's ring.
   (*c*) Tie a tight cotton ligature posterior to Weissmann's ring.
   (*d*) Tie a tight cotton ligature anterior to Weissmann's ring and a second ligature posterior to the ring.
4. Examine the experimental animals at 24 hour intervals. (Keep the larvae at 18–20°C.)
5. Repeat stage 3 with the larvae which has been kept for three days.
6. Examine after 24 hours.

Darkening of the skin indicates the onset of pupation. What conclusions can be drawn from the results of these experiments? Do you feel that further experiments are necessary?

Requirements

Apparatus

Petri dishes
Cotton
Scissors
Refrigerator (optional, but chilled larvae are easier to ligature)

Biological

Blowfly larvae (A stock may be obtained from an angling shop and fed on raw meat or liver in a perforated zinc cage. The adults produced will lay eggs and then larvae of known age will be obtained)

An investigation of the effect of adrenaline and acetylcholine on the heart of a frog

Procedure

1. Pith the frog and pin it on its back in a dissecting dish.
2. Dissect the frog to expose the heart.
3. Carefully remove the pericardium.

4. Pierce the tip of the ventricle with a hook made by bending a fine pin. A long thread is attached to the hook.
5. Remove the heart by cutting through the major vessels.
6. The drawn-out end of a wide tube is inserted into the sinus venosus and tied firmly. If the perfusion tube has a side arm to act as an overflow, the experiment can be carried out at a constant pressure.
7. Immediately the heart is attached, it should be flushed out with Ringer.
8. Attach the thread to the writing lever and record the normal heart beat with a drum speed of 2·5 cm per second.
9. Run a 1 in 10^{-7} solution of adrenaline in frog Ringer into the perfusion tube. Record heart beat.
10. Flush with Ringer and wait until normal heart beats are resumed.
11. Perfuse the heart with a 1 in 10^{-6} solution of acetylcholine in frog Ringer. If no response is observed, carry out perfusion a second time, with acetylcholine Ringer to which 0·004 g per litre of eserine has been added. (The eserine blocks the action of cholinesterase which breaks down acetylcholine very rapidly in the tissues.)

Requirements

Apparatus

Kymograph and paper
Smoking equipment
Writing lever
Dissecting dish
Dissection instruments
Pins
Perfusion tube
Thread

Reagents

Frog Ringer solution
Acetylcholine solution (1 in 10^{-6}) made up in frog Ringer
Adrenaline solution (1 in 10^{-7}) made up in frog Ringer
Acetylcholine solution (1 in 10^{-6}) made up in Ringer with 0·004 g/litre of eserine

Biological

Frog or *Xenopus*

PLANT HORMONES

An investigation of the action of light on the direction of growth of oat coleoptiles

When oat seedlings are stimulated by one-sided illumination, they respond by growing towards the light. Decapitated oat seedlings do not show the same response. This suggests that the tip of the seedling acts as a light receptor. Careful observation of a coleoptile growing towards light will show that the region in which differential growth, or curvature occurs is somewhat distant from the

255

receptor. Clearly there must be some means of communication between the two regions. In 1910, a Danish scientist, Boysen-Jensen, carried out an ingenious experiment which shed some light on the mechanism involved. Later, in 1926, a Dutch scientist, F. W. Went, carried out another experiment which proved beyond doubt that the means of co-ordination between the receptor and response regions was the diffusion of a chemical substance. Further research revealed that indolyl-acetic acid or related compounds are associated with the light curvature response. The experiments of Boysen-Jensen and Went are classic examples of the advances which can be made in scientific knowledge by the use of well designed and controlled experiments of extreme simplicity.

1. Boysen-Jensen's experiment

Procedure

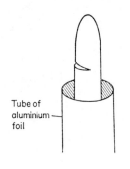

Tube of aluminium foil

1. Using a new razor blade which has had its protective film of grease removed (wipe gently, **away** from the edge of the blade with a piece of cotton wool damped with ethyl acetate, then rinse in alcohol and distilled water), make a wedge-shaped incision on one side of a coleoptile. The incision should be made with the far side of the coleoptile resting against a piece of cork for support. The incision should be approximately half through the coleoptile.
2. When 4 coleoptiles have been prepared in this way they are enclosed with a tube of aluminium foil so that only the tip of the coleoptile will protrude above the tube. (The tubes may be formed by rolling suitable strips of aluminium foil around a piece of thin glass rod or heavy gauge wire.)
3. Set up the seedlings as follows.

   (*a*) One pair in a darkened room with a dry atmosphere, illuminated by a bench lamp from one side. One of the coleoptiles has its incision facing the lamp, the other facing away from the lamp.

   (*b*) The second pair of coleoptiles is set up in a similar way, but is enclosed in a belljar with the base lined with wet blotting paper to give a moist atmosphere.

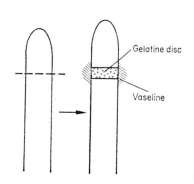

Gelatine disc

Vaseline

4. Prepare a fifth coleoptile by carefully decapitating it, 2–3 mm from the tip. Cut a small piece of gelatine from the Petri dish provided and place the gelatine on the coleoptile and then replace the tip of the coleoptile. Vaseline is then used to ring the join to hold the assembly together and prevent drying out. Place in a darkroom and stimulate with a bench lamp from one side.
5. Examine all seedlings after 2–3 hours.

256

2. Went's experiment

Procedure

1. With a clean sharp razor blade, decapitate several coleoptiles and place the coleoptile tips, cut surface down, on the gelatine in the Petri dish provided.
2. Place the Petri dish in the dark for 1 hour, then remove the coleoptile tip and using a sharp pointed scalpel, cut a tiny cube of gelatine.
3. Decapitate a fresh coleoptile and place the cube of gelatine on it eccentrically. A second coleoptile should be decapitated and left untreated as a control.
4. Place in even illumination and examine after 2–3 hours.

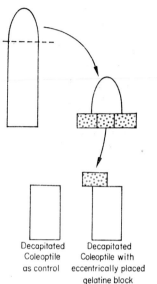

Decapitated Coleoptile as control   Decapitated Coleoptile with eccentrically placed gelatine block

Requirements

Apparatus	Reagents
New, sharp, two-edged razor blades	Gelatine (dissolve 12 g of powdered gelatine in 100 ml water and pour into Petri dishes. Gelatine layer should be 1 mm deep)
Aluminium foil	
Bench lamp	
Belljar	
Blotting paper	Vaseline
Petri dish	
Scalpel	**Biological**
Flower pots	Oat seedlings grown in small pots (6 per pot)

Project

The curvature response to light of different wavelengths differs. Could you design an experiment, using monochromatic filters from the physics department, to show whether the differences are due to difference in reception of the stimulus or to differing degrees of destruction of the plant hormone by light of different wavelength?

An investigation of the effect of indolylacetic acid on the growth of oat coleoptiles

Procedure

1. You are provided with three pots containing oat seedlings. Treat them as follows:

   Pot 1. Label **control** and leave untreated.

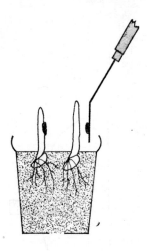

Pot 2. Smear lanolin cream gently on one side of each coleoptile. (Warm the lanolin to 40°C so that it spreads easily—a seeker makes a good applicator.)

Pot 3. Smear the coleoptiles with a lanolin cream containing indolyl-acetic acid.

2. Label pots 2 and 3 to show both the treatment received and which side of the coleoptiles were treated.

3. Place in dark and examine 5–12 hours later.

Requirements

Apparatus

Small flower pots
Labels
Seeker

Reagents

Lanolin cream

Lanolin cream containing indolyl-acetic acid (IAA). This may be purchased ready for use, or an emulsion prepared as follows:

1. Dissolve 10 mg crystalline indolyl-acetic acid in 3 ml of ethanol

2. Add the solution of IAA in ethanol, drop by drop to 100 ml distilled water in a beaker, stirring vigorously. (This gives a 0·01% solution)

3. Place 20 g lanolin in a beaker in a water bath and warm until the lanolin just melts

4. Add 10 ml of the 0·01% IAA solution to the melted lanolin. The addition should be made slowly, stirring vigorously until a creamy emulsion is formed. (If a laboratory emulsifier is available, it may be used)

5. The emulsion will keep for at least a year in a refrigerator

Biological

2 flowerpots each containing 5 or 6 oat seedlings. For experimental work the coleoptiles should be $\frac{1}{2}$–$\frac{3}{4}$ inch high. Several sets of pots may be sown at 3 day intervals to ensure coleoptiles in the right condition on the day required. If germination and growth are more rapid than expected, pots with $\frac{1}{2}$ inch coleoptiles may be placed in the refrigerator to retard them until required

---

The effect of auxins on abscission

When many leaves, flowers and fruits abscisse, a specialised zone of cells forms across the base of the petiole. The intercellular cementing substance breaks down and the cells separate. Only the vascular strand remains intact. Leaf fall then occurs by the mechanical breaking of the vascular tissues. The abscission zone is not formed while there is a good supply of auxin from the leaf, but when the leaf becomes old and this supply decreases, the abscission zone forms.

An effective way of cutting off the auxin supply is to remove the leaf blade.

The knowledge is made use of in agricultural practice, for instance the premature abscission of ripening fruit is prevented by spraying with 2,4-dichlorophenoxyacetic acid or naphthalenc-acetic acid.

In contrast to this the mechanical harvesting of cotton is aided if there are no leaves on the crop, therefore chemicals are applied to kill the leaf blades and thereby reduce the auxin production and so accelerate leaf abscission.

Procedure

1. Starting with the third leaf pair below the apex of a *Coleus* plant, cut the petioles of the leaves at the base of the blade.
2. Repeat this with each of the next two leaf pairs.
3. Apply plain lanolin paste to the cut surface of three petioles.
4. Apply 0·1% IAA in lanolin to the other three petioles.
5. Label the latter three by hanging ring reinforcements on to the petiole.
6. Observe regularly, noting the time at which each petiole falls.

Requirements

Apparatus

Razor blade
Seeker
Ring re-inforcements

Reagents

Lanolin
Lanolin containing indolyl-acetic acid (IAA) 0·1%. Made up as for the earlier experiment but using 100 mg of IAA

Biological

Young *Coleus* in a pot

The inhibition of plant development with anti-auxins

The search for chemicals possessing auxin properties produced some substances which appear to counter-act the effects of auxins. Some will prevent the auxin-stimulated elongation of Avena coleoptiles. Maleic hydrazide will counter-act apical dominance.

Procedure

1. Take 2 6 in high tomato plants in 4 in pots.
2. Label one as the control.
3. Using an atomizer, spray one with 0·4% maleic hydrazide in an aqueous solution. Thoroughly wet all the surfaces of the plant.
4. Observe the plants for 4 weeks.

Questions

1. What is the height of the plant at the end of the fourth week?
2. What development is there of axillary buds?
3. What is the form of the leaves?
4. What is the colour and appearance of the plants?

Requirements

Apparatus

Atomizer

Reagents

0·4% aqueous solution of maleic hydrazide

Biological

2 6 inch tomato plants in 4 inch pots

Appendix 1

Strengths of solutions

Molar solution: (M)

This contains one gram-molecular weight (1 mole) per litre of solution.

A millimolar solution (mM) contains 1/1,000 of the gram-molecular weight per litre.

Example: the molecular weight of HCl is 36·5

Therefore M HCl solution contains 36·5 g/l

Normal Solution: (N)

This contains one gram-equivalent per litre.

Example: the equivalent weight of HCl is the same as the gram-molecular weight.

Therefore a normal solution contains 36·5 g/l.

But H_2SO_4 has a molecular weight of 98 and is bivalent.

Therefore it has an equivalent weight of 49.

Therefore a M-solution contains 98 g/l and a N-solution 49 g/l.

Milli-Moles per litre

The concentration of solutes in biological fluids is often expressed in these terms.

Milli-equivalents per litre

One milli-equivalent is a 1/1,000 of the gram-equivalent weight. This term is used with reference to both molecules and ions.

Example: if the concentration of sodium in blood plasma is 330 mg/100 ml of plasma, the concentration of Na^+ in plasma is 330/23 m-equivalent/100 ml, i.e. 3,300/23 m-equivalent/1 or 143 m-equivalent/l. (The atomic weight of Na is 23 and the Na^+ ion is monovalent, therefore 1 m-equivalent is 23 mg.)

Appendix 2

Units and conversion factors

Mass

gram	g	1 kg	= 1,000 g
kilogram	kg	1 g	= 1,000 mg = 1,000,000 μg
milligram	mg	1 g	= 0·035 oz
microgram	μg	1 oz	= 28·4 g
		1 lb	= 453·6 g = 0·4536 kg
		1 kg	= 2·2 lbs

Volume

Litre	l	1 l	= 1,000 ml
Millilitre	ml	1 ml	= 1,000 μl
Microlitre	μl	1 l	= 1·76 pints
cubic centimetre cm³		1 l	= 35·196 fluid oz
		1 pt	= 568 ml
		1 oz	= 28·4 ml
		1 gallon	= 4·55 litres

Length

metre	m
centimetre	cm
millimetre	mm
micrometre	μm
nanometre	nm
Ångstrom unit	Å
picometre	pm

$1 \text{ m} = 100 \text{ cm} = 1{,}000 \text{ mm} = 10^6 \mu\text{m} = 10^9 \text{ nm} = 10^{10} \text{ Å}$
$= 10^{12} \text{ pm}$

1 cm	= 0·394 in
1 kilometre	= 0·62 miles
1 in	= 2·54 cm
1 mile	= 1·61 kilometres
1 ft	= 30·48 cm
1 yd	= 0·9144 m

Temperature

Conversions from the Fahrenheit scale to the Celsius scale may be made with the aid of the following formula:

$$9/5 \text{ C} = \text{F} - 32$$

where C is the temperature in degrees Celsius and F is the temperature in degrees Fahrenheit.

Appendix 3

Buffer solutions

Universal buffer mixture

This can be purchased from British Drug Houses. It consists of a mixed salt. If the contents of the purchased tube are dissolved to make 1 litre of solution, the pH value of this solution will be 3·1. Buffer solutions with pH values ranging from 2·7 to 11·4 can be made from this by the simple addition of 0·2 N HCl or NaOH. When these additions have been made, the pH value will be $3·1 \pm 0·1185 \; V$ where $V =$ the number of ml of 0·2 N HCl or NaOH added to 100 ml of solution with the subsequent dilution to 200 ml with water. The NaOH being (+) and the HCl being (−).

De-ionised or freshly distilled water must be used, i.e. the addition of each ml of HCl or NaOH will change the pH by ± 1·185 respectively.

pH range 2·2-8·0

To make up 100 ml of buffer add 0·2 M Na_2HPO_4 to 0·1 M citric acid in the proportions shown.

pH	Na_2HPO_4	Citric acid	pH	Na_2HPO_4	Citric acid
2·2	2·00	98·00	5·2	53·60	46·40
2·4	6·20	93·80	5·4	55·75	44·25
2·6	10·90	89·10	5·6	58·00	42·00
2·8	15·85	84·15	5·8	60·45	39·55
3·0	20·55	79·45	6·0	63·15	36·85
3·2	24·70	75·30	6·2	66·10	33·90
3·4	28·50	71·50	6·4	69·25	30·75
3·6	32·20	67·80	6·6	72·75	27·25
3·8	35·50	64·50	6·8	77·25	22·75
4·0	38·55	61·45	7·0	82·35	17·65
4·2	41·40	58·60	7·2	86·95	13·05
4·4	45·10	55·90	7·4	91·85	9·15
4·6	46·75	53·25	7·6	93·65	6·35
4·8	49·30	50·70	7·8	95·75	4·25
5·0	51·50	48·50	8·0	97·25	2·75

pH 3·0-6·2 Citrate buffer

Make up a 0·1 M solution of citric acid (19·21 g per 1,000 ml) and a 0·1 M solution of sodium citrate (29·41 g per 1,000 ml).

To make the appropriate buffers add the quantities of each of these solutions indicated below and make the solution up to 100 ml with de-ionised water.

Citric acid	Sodium citrate	pH	Citric acid	Sodium citrate	pH
46·5	3·5	3·0	23·0	27·0	4·8
43·7	6·3	3·2	20·5	29·5	5·0
40·0	10·0	3·4	18·0	32·0	5·2
37·0	13·0	3·6	16·0	34·0	5·4
35·0	15·0	3·8	13·7	36·3	5·6
33·0	17·0	4·0	11·8	38·2	5·8
31·5	18·5	4·2	9·5	40·5	6·0
28·0	22·0	4·4	7·2	42·8	6·2
25·5	24·5	4·6			

pH 3·6-5·6 Acetate buffer

Make up a solution of acetic acid (0·2 M, 11·55 ml in 1,000 ml) and a solution of sodium acetate (0·2 M, 16·4 g of CH_3COONa per 1,000 ml or 27·2 g of $CH_3COONa, 3H_2O$ per 1,000 ml).

Add the amounts of these indicated below and make the solution up to 100 ml.

Acetic acid	Sodium acetate	pH	Acetic acid	Sodium acetate	pH
46·3	3·7	3·6	20·0	30·0	4·8
44·0	6·0	3·8	14·8	35·2	5·0
41·0	9·0	4·0	10·5	39·5	5·2
36·8	13·2	4·2	8·8	41·2	5·4
30·5	19·5	4·4	4·8	45·2	5·6
25·5	24·5	4·6			

pH 5·8-8·0 Phosphate buffer

Make a solution of sodium dihydrogen phosphate (31·2 g of $NaH_2PO_4, 2H_2O$ per litre) and a solution of disodium hydrogen phosphate (28·39 g of Na_2HPO_4 per litre or 71·7 g of Na_2HPO_4, $12H_2O$ per litre). Each of these solutions will be 0·2 M solutions. To make up the required buffers, add the indicated amounts of the solutions together and make up to 100 ml with de-ionised water.

NaH_2PO_4	Na_2HPO_4	pH	NaH_2PO_4	Na_2HPO_4	pH
46·0 ml	4·0 ml	5·8	19·5 ml	30·5 ml	7·0
44·0	6·2	6·0	14·0	36·0	7·2
40·7	9·2	6·2	9·5	40·5	7·4
36·7	13·2	6·4	6·5	43·5	7·6
31·2	18·7	6·6	4·2	46·7	7·8
25·5	24·5	6·8	2·6	47·3	8·0

pH 7·6-9·2 Boric acid-Borax buffer

Make up a solution of boric acid (12·4 g per litre) and one of borax (19·05 g per litre). The acid solution is 0·2 M and the borax 0·5 M (0·2 M in terms of sodium borate).
To 25 ml of the acid add the indicated amount of borax solution and make the solution up to 100 ml with de-ionised water.

Borax solution	pH	Borax solution	pH
1 ml	7·6	8·7	8·6
1·55	7·8	15·0	8·8
2·45	8·0	29·5	9·0
3·6	8·2	57·5	9·2
5·7	8·4		

pH 9·2-10·6 Carbonate-bicarbonate buffer

Make up a solution of anhydrous sodium carbonate (21·2 g per litre), and a solution of sodium bicarbonate (16·8 g per litre).
Both these solutions will be 0·2 M solutions.
To make up the required buffers, add the indicated amounts of the solutions together and make the solution up to 100 ml with de-ionised water.

264

Na_2CO_3	$NaHCO_3$	pH	Na_3CO_3	$NaHCO_3$	pH
2·0 ml	23·0 ml	9·2	13·7	11·2	10·0
4·7	20·2	9·4	16·5	8·5	10·2
8·0	17·0	9·6	14·2	5·7	10·4
11·0	14·0	9·8	21·2	3·7	10·6

Appendix 4

Isotonic saline solutions

A number of experiments require the bathing of tissues in a solution which is isotonic with the body fluid of the animal. For temporary microscopical preparations and short experiments, an isotonic solution of sodium chloride may be used. For more prolonged physiological investigations it is usually preferable to use one of the more comprehensive Ringer solutions. For simplicity of nomenclature, these more carefully balanced solutions are referred to as Mammal Ringer, Insect Ringer, etc, although the formulation given may be a more recent modification of the original Ringer's formula. In some experiments a trace of glucose is added to the Ringer solution to provide a respiratory substrate.

Saline solutions

These are sometimes referred to as 'normal salines'—a term which should never be used as it is easily confused with normal solutions in the chemical sense.

For invertebrate tissues 0·75% NaCl made up in de-ionised water
For amphibian tissues 0·64% NaCl ,, ,, ,, de-ionised ,,
For mammalian tissue 0·9% NaCl ,, ,, ,, de-ionised ,,
For mammalian blood 0·6% NaCl ,, ,, ,, de-ionised ,,

Ringer's solutions

For frog Grams per litre
NaCl 6·5
$CaCl_2.6H_2O$ 0·12
KCl 0·14
$NaHCO_3$ 0·20

For mammal
NaCl 8·0
$CaCl_2.6H_2O$ 0·2
KCl 0·2
$NaHCO_3$ 1·0
$MgCl_2.6H_2O$ 0·1
NaH_2PO_4 0·05

If required, 1 gram of glucose per litre may be added.

Grams per litre

For locusts and other insects

NaCl	7·6
CaCl$_2$.6H$_2$O	0·22
KCl	0·75
MgCl$_2$ 6H$_2$O	0·19
NaHCO$_3$	0·37
NaH$_2$PO$_4$	0·48

For earthworm

NaCl	6·0
CaCl$_2$.6H$_2$O	0·2
KCl	0·12
NaHCO$_3$	0·1

For 'Astacus'

NaCl	12·0
CaCl$_2$.6H$_2$O	1·5
KCl	0·4
MgCl$_2$.6H$_2$O	0·25
NaHCO$_3$	0·2

For marine crustaceans including 'Carcinus'

NaCl	31·0
CaCl$_2$.6H$_2$O	1·37
KCl	0·99
MgCl$_2$.6H$_2$O	2·35
NaHCO$_3$	0·22

For chick embryos

NaCl	7·0
CaCl$_2$.6H$_2$O	0·24
KCl	0·42

Appendix 5

pH of some fluids of biological interest

Seawater	8·0–8·1
Acid swamps	3·5–5·5
Lakes and rivers	6·5–9·0
Marine invertebrates	7·2–7·8
Insects	6·0–8·8
Drosophila	7·1
Helix	7·8–8·4

Man: Blood	7·3–7·5
Saliva	6·5–7·5
Sweat	3·8–6·5
Bile	6·8–7·0
Urine	4·8–8·4
Stomach contents	1·0–3·0
Duodenal contents	4·8–8·0

Very little reliable data on the pH of biological material has been published and even less appears to be known about the fluctuations which occur within a species. (With the exception of urine and blood of man, where fluctuations have a diagnostic importance in medicine.)

Appendix 6

Oxygen consumption of some common animals at room temperature

Amoeba	0·2 ml/g/hr
Paramecium	0·5
Sea anemone	0·01
Earthworm	0·06
Crayfish	0·07
Butterfly resting	0·6
Butterfly flying	100·0
Blowfly larva	1·3
Mouse resting	2·5
Mouse running	200·0
Man resting	0·2
Man working hard	40·0

Appendix 7

The culture of laboratory animals

The Cruelty to Animals Act (1876). Experiments on vertebrates are rigorously controlled by the act and every care should be taken not to infringe this rather complicated piece of legislation. Care has been taken to select experiments for this book which do not infringe the act. Any experimental work about which the teacher has doubt should be checked by seeking the advice of the Home Office Inspector for his area, whose address is obtainable from the Under Secretary of State, Home Office, London, S.W.1.

A licence is required for any experiment which causes harm to a vertebrate or for an experiment in which the results are unknown. Injections for experimental purposes are legal only if the results are known and the animal suffers no harm, e.g. the injection of Pregnyl in reasonable doses, to induce spawning in Xenopus, is legal.

The pithing of frogs is regarded as a means of killing and a pithed frog is regarded as a dead animal within the meaning of the Act. Hence kymograph experiments using pithed frogs are legal. **The authors feel that such experiments should be kept to an absolute minimum and that pithing should never be carried out in the presence of a class. The teacher has an obligation not to act in any way which might encourage a callous attitude on the part of the student.**

General notes on the care of laboratory animals

Cleanliness

If laboratory animals are to be kept in a healthy condition, it is essential to keep the animal room scrupulously clean and free from unnecessary equipment—it should not be used as a store. Cages should be cleaned weekly and all litter swept up. It is advisable to spray or wash all floors and benches with a suitable disinfectant such as Tego MGH at regular intervals. This precaution will very considerably reduce the incidence of respiratory infections, intestinal infections and skin conditions. A very useful leaflet *Animal House Disinfection*, is available from the distributors of Tego, Messrs. Hough, Hoseason and Co. Ltd., Chapel Street, Manchester, 19.

Ventilation, Humidity and Light

Fortunately small mammals are tolerant of considerable differences in the physical environment and this makes it possible to house a number of species in the same room. The ventilation system should allow 6–10 changes of air per hour without exposing the animals to draughts. Good ventilation plays an important role in reducing respiratory infection. Optimum humidity varies from 45% in the case of rabbits, to 65% for mice. For a school animal house any figure between these points will be found to be suitable for a mixed stock.

Lighting conditions are not critical but should be diurnal. Seasonal variation in litter frequency is related to day length.

Cages

Make sure that the cages are appropriate to the size of the animal. Freedom of movement and room to stretch out fully are essential for health.

Wire cages are easy to sterilise with Tego but are much more expensive than wooden boxes which are preferred by many large scale users. For school use polypropylene mouse and rat cages with wire tops incorporating a food hopper and water bottle are ideal.

Whatever type of cage is chosen, the floor should be lined with at least $\frac{1}{2}$ inch of sawdust which should be changed each week. If possible use sterile sawdust obtained from animal foodstuff suppliers as the untreated sawdust obtained from a local sawmill is often contaminated by the faeces of cats, dogs, mice and rats and can be an unexpected source of parasite and bacterial infections. When wire cages are used it is most important to see that the floor is covered by sawdust to prevent the development of sores on the feet of the occupants. Pregnant animals should be given a good supply of wood shavings for nesting. If rabbits are kept, hay is the most suitable nesting material.

Cages should be clearly labelled or numbered in such a way that the animals are unable to chew the cards.

Food and water

A balanced diet in the form of pellets is available to suit the needs of most mammals and can be ordered through a local corn-merchant. Mashes are satisfactory but are much more labour-consuming than pellets. When working on a tight budget, daily feeding is more economical than weekly filling of a hopper. If animals are kept on a minimal diet litter frequency and size may be reduced.

Water should always be available, independently of the food supply. If there is a considerable fluctuation in the air temperature, difficulty may be experienced with the inverted type of water bottle as expansion of an air bubble may force water into the cage. Water bottles should be changed and disinfected regularly.

Disease

The diagnosis and treatment of disease is a job for the expert and in a school animal house should not normally be attempted. Ailing animals and those with lesions should be humanely killed and the cage sterilised. If this precaution is taken and animals are only obtained from accredited breeders, epidemics should not occur. Animals should never be introduced from pet shops.

Noise

Mice and rats are seriously affected by high-frequency noise such as the emptying of dustbins and hammering.

Killing

The following methods are humane and are suitable for use in the school:

1. Breaking the spinal chord. The animal should be held by the hind legs and its cervical region cracked sharply across the edge of a bench.

269

2. An overdose of chloroform, nitrogen or coal gas may be administered. Ether should not be used on young mice or rats as they may recover from its affects some time after apparent death.

Chloroform should never be administered in the animal house as male mice are very susceptible to the vapour and may die from degeneration of the liver 2–3 days after exposure.

3. Place the animal in a strong polythene bag and fill this with CO_2 from a cylinder. It may be convenient to place the cage and the animal in the bag. This method is recommended by the Universities Federation of Animal Welfare.

As far as possible, animals should not be killed in the presence of a class.

Mongolian Gerbil

Oestrous cycle	Has not yet been precisely defined but would appear to be about 4–7 days in length
Gestation	24–28 days
Weaning age	24 days
Mating age	10–12 weeks
Litter frequency	Post-partum mating may take place and hence litters may be produced monthly
Litter size (average)	5–7

Breeding

They are best kept in monogamous pairs and the males can be left with the female throughout the breeding cycle.
There is a tendency for some pairs to cannibalise their young. Several factors, such as dietary deficiency and inbred stock, seem to cause this. It is best to leave the litter completely undisturbed for the first few days.
Bedding should not be changed for about two weeks.
Juvenile animals may be kept together in large groups of the same sex and it is possible to continue keeping them in this manner when the animals are adult. If two strange adults are placed together they will fight and severely injure each other.

Mice

Oestrous cycle	4–5 days
Gestation	19–21 days
Weaning age	21 days
Mating age	6–8 weeks
Litter frequency	8–12 yearly
Litter size (average)	7–8
Optimum temperature	21°C
Humidity	50–65%

Breeding

Although mice may mate at 6 weeks, the usual age in controlled breeding programmes is 8–12 weeks. If mice are kept in polygamous trios, that is, one male and two females per cage, litters may be expected every 3–4 weeks. Young mice should be weaned before the next litter appears. If mice from a single stock are inbred for 7 or 8 generations a fall in average litter size may occur.

Rats

Oestrous cycle	4–5 days
Gestation	21–23 days
Weaning age	28 days
Mating age	10–12 weeks
Litter frequency	7–9 yearly
Litter size (average)	7
Optimum temperature	21°C
Humidity	45–55%

Breeding

The animals should be 10–12 weeks old. Mating is usually carried out in polygamous trios. Females, when pregnant, should be removed to a separate cage with plenty of nesting material. The female and litter should be left undisturbed as far as possible until weaning. After weaning, the young should be kept in a high density population for at least a week.

Rabbits

Oestrous cycle	No definite cycle. Ovulation occurs 10 hours after mating. There is evidence of rhythmic sexual activity.
Gestation	28–31 days
Weaning age	6–8 weeks
Mating age	6–9 months
Litter frequency	4 yearly
Litter size (average)	4
Optimum temperature	18°C
Humidity	40–45%

Breeding

9-month-old rabbits should be mated when the female is on heat. It is advisable to introduce the doe to the buck's cage, not vice versa. The vulva of the doe becomes red, moist and swollen when she is on heat. After 24 days the doe should be transferred to a large cage with a screened breeding compartment. Plenty of hay for nesting must be provided and the litter should not be disturbed for 10 days if cannibalism is to be avoided. The young should be weaned when they are 6–8 weeks old and the sexes separated into different runs.

Guinea-pigs

Oestrus cycle	16 days
Gestation	Usually around 63 days but may vary between 59–72 days
Weaning age	14–21 days
Mating age	12–20 weeks
Litter frequency	3 yearly
Litter size (average)	3
Optimum temperature	21°C
Humidity	45%

Breeding

12 weeks is the usual minimum age. One male is placed in a pen with 5–10 females allowing at least 1 square foot of floor space per animal. Pregnant females are placed in separate cages.

Food

In addition to pelleted food, guinea-pigs should be given 2 ounces of fresh food daily—carrots, cabbage and swedes are suitable.

Hamsters

Oestrous cycle	4–5 days
Gestation	16–17 days
Weaning age	3–4 weeks
Mating age	7–9 weeks
Litter frequency	3 or 4 yearly
Litter size (average)	6
Optimum temperature	21°C
Humidity	40–50%

Cages

Hamsters must be kept in strong cages, they will very quickly gnaw their way out of wood, aluminium or zinc cages. The polypropylene cages manufactured by North Kent Plastic Cages Ltd, Home Gardens, Dartford, Kent, are excellent. Hamsters must be provided with wood to gnaw otherwise their incisor teeth will overgrow until feeding is not possible.

Breeding

Mating should not be carried out until the female is 12–20 weeks old. Mating should be carried out on a table top under supervision as the female is liable to attack the male savagely after copulation or if she is not receptive. The female comes on heat every fourth night, usually after sunset. If a female is on heat she will adopt a mating position when placed with the male.

Pelleted diets

For all school purposes the following are suitable:

Mice, Rats and Mongolian gerbils	Diet FFG(M) or Diet 86

Rabbits and Guinea-pigs Diet 18
Hamsters   Diet FFG(M) or 86 or 18 with some green supplement if possible.

Diet RGP is suitable for rabbits, hamsters and guinea-pigs without a green supplement. E. Dixon and Sons (Ware) Ltd., Crane Mead Mills, Ware, Herts. or Peter Fox (Scientific Animals) Ltd. Home Farm, Aldenham Park, Elstree, Herts., supply all the above diets in $\frac{1}{2}$ cwt sacks.

Xenopus

The African clawed toad has many advantages as a laboratory animal when compared with *Rana* and *Bufo*. Stocks are easily maintained in the laboratory and breeding can be carried out at any time of the year. *Xenopus* will not breed until they are two years old—it is therefore necessary to order a breeding pair when starting a colony.

Containers

Ordinary aquaria or plastic tanks are suitable. Use a large size, 20 in × 16 in × 6 in deep for stocks and a smaller size, 12 in × 8 in × 8 in for breeding. Plastic tanks of these dimensions are available from Messrs Thermo Plastics Ltd, Luton Road, Dunstable. The tanks should be firmly covered with a transparent sheeting, reinforced with cross wires, such as Claritex which is available from most ironmongers.

Temperature

Adult stocks can be maintained without heat as long as the ambient temperature does not fall below 10°C. For breeding and rearing the larvae, the tank should be heated to 23°C using an aquarium heater and thermostat.

Feeding

Earthworms are an ideal food for the adults as they are consumed without fouling the water. If earthworms are not available, finely chopped liver or heart may be given. The stock should be fed twice weekly, and the water changed immediately afterwards. Changing the water is facilitated if the plastic tanks are fitted with taps.

Diseases

2% mortality per year is to be expected in a large stock kept under clean conditions with no overcrowding. Adults should be kept at a population density of 10–15 *Xenopus* to 30 litres of water.

Marking

Xenopus may be marked by clipping one or more claws.

Breeding

Hormonal stimulation is necessary to ensure breeding in captivity. 2 cm³ disposable syringes fitted with No. 18 needles are used and both male and female toads are given a primer and a final injection.

	Primer	Final
Female	100 i.u. Pregnyl	300 i.u. Pregnyl
Male	50 i.u. Pregnyl	100 i.u. Pregnyl

A breeding pair are kept at 23°C for 4 weeks before the primer injection is given. The final injection is given four days after the primer. Pregnyl is obtainable from Organon Ltd, Crown House, Morden, Surrey and is supplied in separate ampoules of Pregnyl and distilled water. The water is taken into the syringe and transferred to the ampoule of Pregnyl which dissolves instantly and can be drawn up into the syringe. **Air bubbles are removed by holding the syringe, needle uppermost, and pushing the plunger in until all air is expelled.** The injection is made by inserting the needle into the thigh, and passing it forward in the subcutaneous connective tissue to the dorsal lymph space which lies just beneath the skin, between the 'stitch marks' and the mid-line.

If, after injection the toads pair without spawning the temperature should be raised to 28°C.

After spawning, the adults should be removed and the eggs distributed between containers so that each egg is in direct contact with the surrounding water and no eggs are left enclosed in solid masses. Gentle aeration may be given by means of an aquarium pump fitted with a diffuser block.

The newly-hatched tadpoles may be fed on dried nettle powder which can be obtained from health food stores. Metamorphosis occurs approximately 7 weeks after hatching. During the metamorphosis period the larvae may be fed on micro-worm, *Daphnia*, *Tubifex* or white-worms. After metamorphosis, finely-shredded liver and heart may be given.

Xenopus in all stages are available from Phillip Harris Ltd.

Locusts

Locusts are very useful laboratory animals for schools as they can be used for a wide range of practical work. They are relatively easy animals to maintain as a permanent stock.

Cages

For stock colonies and breeding the most suitable cage is a glass fronted container approximately 15 in × 15 in × 20 in high, with a false floor of perforated zinc. The cage should be heated by means of two electric light bulbs, one below the false floor and one in the locust compartment. The false floor should be arranged so that containers of sand for egg laying may be placed underneath. The minimum size for the egg containers should be 4 in deep by 1¼ in diameter. An excellent cage of the pattern used by the Anti-Locust Research Centre, complete with egg tubes is available cheaply from Phillip Harris Ltd. A cage of this type may be used to house up to 200 adults.

For raising locusts in isolation, an excellent container may be made by replacing the top of one of the new pattern Kilner jars with a disc of perforated zinc. The disc is best attached with Evo-Stik or Araldite. Several containers of this type may be heated by placing them around a 40 watt bulb at a distance of approximately 8 inches.

Cylindrical perspex or cellulose acetate cages 15 in × 6 in are available from P. K. Dutt Ltd, at 17s. 6d. and may house 20–30 adults.

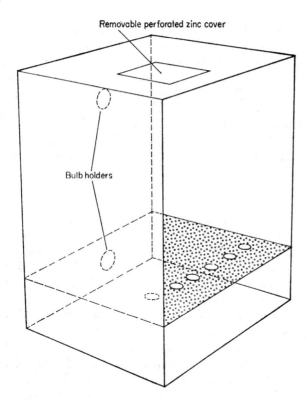

Removable perforated zinc cover

Bulb holders

Perches

It is essential to provide each cage with twiggy branches or a cylinder of large mesh wire netting to provide perching space. Unless the instars can hang freely when moulting a high proportion of deformed adults will occur.

Temperature

Locusts will tolerate a wide range of temperatures and can be kept in an unheated laboratory during the summer, development will be slow and mortality high under these conditions. Ideally locusts should be kept within the range 28–34°C. Adjustments may be made by changing the bulbs in the cages. A suitable arrangement is a 25 watt bulb below the floor and 40 watt bulb in the locust compartment. A temperature gradient will occur in the cage and provided that plenty of perching space has been given, the locusts will bask in the optimum region.

Humidity

Excess humidity will favour disease. If condensation appears on the cages, or faeces are soft and wet instead of hard and dry, steps should be taken to lower the humidity. Assuming that the temperature of the cage is not too low, the most usual cause of excessive humidity is overfeeding with lush grass.

Feeding

Locusts may be fed entirely on grass. Sufficient fresh grass should be placed in the cage each day, the amount should be only very slightly more than can be consumed in the 24 hours before the next feed. Overfeeding makes cleaning unnecessarily difficult and raises the humidity. Grass need not be given daily if a plentiful supply of wheat bran, available from pet shops, is always present in the cage. Wheat bran is particularly useful in schools for overcoming the weekend feeding problems. When grass is not available, cabbage may be used as a substitute but has the disadvantage of resulting in very malodorous faeces.

Breeding

The female lays eggs in a frothy 'pod' in a hole which she excavates in moist sand. Each pod contains 30–100 eggs, each of which are the size of a rice grain. For laboratory use, sharp builders' sand should be sieved to remove particles over $\frac{1}{16}$ of an inch, then washed with repeated changes of water until all dust-like particles have been removed. The sand should then be sterilised in an oven or autoclave. Once a stock of sand has been prepared it can be used repeatedly, provided that it is resterilised before each use. To obtain the correct moisture content, the dry, sterilised sand is mixed in the proportion of 100 parts sand to 15 parts of water by volume. The sand is then placed in containers at least 4 inches deep, beneath holes in the floor of the cage. Porous containers should not be used as they dry out too rapidly. The containers should be replaced every one or two days, even if they have not been used. Gently tip off the top $\frac{1}{2}$ inch of dry sand to check for the presence of pods. Two or three pods in a standard tube are sufficient—if the numbers are greater than this there will be a sharp fall-off in the hatching percentage. After checking for the presence of pods, the containers should be loosely covered to prevent excessive evaporation and set to incubate at 28–32°C. At these temperatures *Locusta migratoria migratorioides* (African migratory locust) will

276

hatch in 11–16 days, and *Schistocerca gregaria* (Desert locust) in 12–17 days.

Five hundred first stage hoppers may be kept in a standard cage and will produce 100–200 adults. Although the mortality rate is high, few corpses will be observed as cannibalism is prevalent.

Locust allergy

Some people, after prolonged exposure to high density populations of locusts, show allergy symptoms. The severity of the symptoms and the rapidity of onset depends on the degree of exposure to locusts and the individual's susceptibility to allergic conditions. Any person with a family history of allergies, hay-fever or asthma should avoid regular handling of locusts or working for long periods in rooms where locusts are kept in large numbers. From the school point of view, the only people at risk are technicians responsible for feeding and cleaning, and staff who spend most of their time working in the same laboratory. At present, expert opinion suggests that there is no risk for children who only use the laboratory once or twice per week, but that it would be unwise to keep locusts in a classroom used by one class for long periods unless the locusts were few in number. The authors prefer to keep their locusts in a corridor where they are a source of considerable interest without exposing any individual to prolonged contact.

The symptoms of allergy start with an apparent head cold and sneezing, these symptoms often disappearing at night. Concurrently, or sometimes at a later date, a characteristic skin irritation of the hands and fore-arm. If these symptoms appear, the individual should stop working with locusts and the symptoms will disappear without trace. Once a person is sensitised to locusts, continued exposure may lead to respiratory conditions such as bronchitis and asthma.

Amoeba

Shallow, glass containers capable of taking a 2 cm deep culture solution should be used. The tops of household pyrex dishes are ideal. Extreme cleanliness is essential to success; all glassware should be washed in Teepol and then rinsed with at least twenty changes of clear tap water, finally rinse in glass distilled water (metal contaminants are lethal).

Culture medium

Chalkley's medium is excellent. It is prepared as a stock solution and diluted for use.

Chalkley's medium (Stock solution)

NaCl	16 g
$NaHCO_3$	0·8 g
KCl	0·4 g
$NaHPO_4$	0·2 g

Make up to one litre with glass distilled water. De-ionised water contains phenols which are harmful to Protozoa.

For use, take 5 ml of stock solution and dilute to 1 litre with glass distilled water. Distribute the medium to the culture dishes, giving each a depth of 2 cm. Boil wheatgrains and place 4 in each dish to grow mould and bacteria and finally *Colpidium* which forms the food for Amoebae.

Technique of culturing

Inoculate the culture with *Amoeba* using Pasteur pipettes for handling. Growth is erratic at low temperatures.

At 7°C there is 1 division per week.

At 30°C division is most rapid but the culture produces small, underfed individuals.

17–23°C is the most suitable range (many workers keep their cultures at 20 ± 1°C).

At 17°C the lifespan is 2–3 days and approximately 2% of the culture are at the division stage at any time.

If dividing *Amoebae* are transferred they may produce binucleate individuals. After inoculating the culture in a Pyrex dish with about 300 *Amoebae*, it should be covered and left for 4 weeks without disturbance. A harvest of 5,000 cells may be expected. The culture should be kept in a fairly dark place to avoid the growth of algae.

Always transfer attached individuals, floating forms are not healthy.

Mass culture

If a large plastic sandwich box is charged with 2 cm depth of Chalkley's medium and inoculated with a 4 week culture (approximately 5,000 cells), it is possible to harvest 5 ml of packed cells (1 million) after centrifuging. For mass culture the *Colpidium* required for food should be cultured separately to ensure a sufficient supply (see below).

Notes

1. Wheat obtained from seed merchants may have been dressed with a mercuric fungicide and is unsuitable for culture media. (Undressed wheat is available from Carters Seeds Ltd and health food stores.) If undressed wheat cannot be obtained, polished rice grains may be substituted.
2. Aromatic compounds must be kept out of culture rooms. Xylene, methyl benzoate and the solvents used in paints are particularly lethal to protozoan cultures.
3. It is almost certainly a waste of time to attempt to collect the proteus group of *Amoebae* from the wild.

Colpidium

Colpidia will usually appear in a suitable culture medium if it is left exposed to the air. An excellent medium is 1% proteose peptone (available from Oxoid or Difco). If cultures are available, the medium should be sterilised and inoculated with *Aerobacter aerogenes*. 24 hours later the culture should be cloudy with bacteria and ready for inoculation with *Colpidium*. Another excellent medium is 0·1% milk in Chalkley's medium, when the culture clears it is at its peak and the *Colpidia* may be removed by means of a centrifuge. (2,000 r.p.m. for 8 minutes will bring down the ciliates, leaving the bacteria in suspension.) If the ciliates are required to feed *Amoebae*, decant off the supernatant liquid, rinse with Chalkley's medium and re-centrifuge to remove the peptone or milk.

Stentor

Stentor grows well in Chalkley's medium. Feed with *Colpidium* every 2 or 3 days and keep the culture in good light. Beakers are suitable vessels.

Paramecium

Add 30 grains of boiled wheat to a litre of Chalkley's medium and inoculate with *Bacillus subtilis*. 24 hours later inoculate with *Paramecium*. 500 ml flasks containing 300 ml of medium, plugged with cotton wool, are ideal vessels. Subculture every 3 weeks.

Algae and green flagellates

Add 10 grains of wheat to 500 ml Chalkley's medium and auto-clave for 15 mins at 15 lb/sq inch pressure. Inoculate from purchased culture and place in a sunny window. The culture should be plugged with sterile cotton wool to keep out airborne spores.

Hydra

Hydra present two main difficulties when kept as cultures. Firstly, they require suitable pond water or a chemically defined medium. Tap water is usually lethal as *Hydra* are sensitive to the copper ions present. Tap water can be used if the copper ions are first chelated by the addition of 50 mg/1 'Versene'. (Di-sodium ethylenediamine tetra-acetate.) The second difficulty is the pro-vision of a suitable supply of small crustaceans for food. Unless young *Daphnia* are readily available, the most convenient food will be newly hatched brine-shrimps (*Artemia*).

Chemically defined medium

Stock solution A.	Potassium chloride	3·75 g
	Sodium hydrogen carbonate	42·00 g
	2-amino-2-(hydroxy-methyl) propane-1 : 3-diol.	60·57 g
	Make up to 1 litre with de-ionised or glass distilled water.	

279

Stock solution B.   Calcium chloride   41·6 g
                             Magnesium chloride (hydrous)   10·2 g
                             Make up to 1 litre with de-ionised
                             or glass distilled water.

The working solution is made up by adding 20 ml of each stock solution to 10 litres of de-ionised or glass distilled water, approximately 9 ml N hydrochloric acid is then added to bring the pH within the range 7·5–7·8.

2-amino-2-(hydroxy-methyl) propane-1 : 3-diol is available from Harrington Bros Ltd, Weir Road, Balham, London S.W.12. It is also marketed under the name Tris Buffer.

Culture technique

Shallow culture dishes such as domestic Pyrex lids are filled with 1 cm depth of medium and kept at room temperature. *Hydra viridis* should be kept in good light, but for *Hydra fusca* a more shady position is to be preferred. *Hydra* should be fed twice per week with newly hatched *Daphnia* or *Artemia*. One hour after feeding, the culture dish should be gently tilted to pour off the medium and any surplus food and refilled with fresh medium.

Sexual reproduction may be induced by placing the culture in a refrigerator for 2–3 weeks at 10–15°C. The culture should be fed normally.

Artemia

Artemia (Brine-shrimp) eggs are available from aquarium shops. They are hatched by placing them in a salt solution at room temperature. At 15–20°C, hatching occurs in 2–3 days. If supplies are required at short notice, they will hatch in 24 hours at 30°C. The culture solution is made by dissolving 360 g sodium chloride in 1 litre of hot water and cooling. For use the stock solution is diluted with 10 parts water. *Artemia* must be rinsed in clean water to remove salt before feeding to *Hydra*.

Appendix 8

Microbiological stains and reagents

Potato dextrose agar

Wash 250 g of potatoes thoroughly and cut into small pieces.
Place them in a muslin bag and steam for 1 hour by suspending it over a pan of boiling water or by placing it in an autoclave.
After steaming, allow the liquid to drip from the bag until drainage is complete. Do not squeeze it.
Make the extract up to 1,000 ml with tap water and add 20 g of glucose.

Add 20 g of agar slowly while warming the extract until it has all dissolved.

Place the agar in the autoclave or pressure cooker to sterilise it.

Egg saline medium

Add 75 ml of beaten egg to 25 ml of a sterile 0·85% solution of sodium chloride.

All the equipment used must first be sterilised.

The eggs must not be more than 4 days old.

Wash the eggs in warm water with a brush and soap. Rinse in running water for 30 minutes. Drain off the water and dry the eggs by washing in methylated spirits, and flame the surface of the egg.

Before handling the clean, dry egg, wash your hands thoroughly. Crack the eggs with a sterile scalpel and pour the contents into a sterile beaker. Beat the eggs with a sterile whisk.

The medium is then poured into McCartney bottles and solidified by placing on a slope in an oven at between 75–85°C for an hour. If the medium has not been prepared with aseptic precautions, it must now be sterilised in an autoclave at 121°C for 15 minutes. For this the screw caps must be tightened to prevent the slant being disrupted by bubbles of steam.

Litmus milk medium

Fresh milk should be steamed for 20 minutes and then left to stand for 24 hours to allow the cream to separate.

Siphon off the milk and add the litmus solution. Add 2·5 parts of litmus solution to 100 parts of milk.

Pour 5 ml amounts into McCartney bottles and sterilise.

If large quantities are being stored, do not add the litmus until about to use the medium as the colour fades on storing.

To make up the litmus solution, grind up 80 g of litmus granules and place them in a flask with 150 ml of 40% ethyl alcohol. Boil this for 1 minute.

Decant off the fluid and add another 150 ml of ethyl alcohol. Boil this for 1 minute.

Decant off the fluid and add it to the first extract.

Make it up to 300 ml with 40% ethyl alcohol.

Add N HCl, drop by drop until the colour is purple. Shake the flask between the addition of each drop.

Robertson's cooked meat medium

Mince 500 g of fresh bullock's heart and place in alkaline boiling water. Use 500 ml of water and 1·5 ml of N NaOH.

Simmer for 20 minutes to neutralise the lactic acid.

Drain off the liquid through muslin while still hot.

Still while hot, press the meat in cloth and dry by spreading it out on filter paper.

Then place in McCartney bottle to a depth of 1 inch.

281

Prepare the broth by adding 500 ml of the liquid filtered from the cooked meat to 2·5 g of peptone and 1·25 g of NaCl.

Steam at 100°C for 20 minutes.

Add 1 ml of pure HCl and filter.

Adjust the pH of the filtrate to 8·2 and steam at 100°C for 30 minutes.

Finally adjust the pH to 7·8.

Cover the meat in the bottles with 10 ml of broth.

Autoclave at 121°C for 20 minutes.

(This medium is suitable for growing anaerobes on in air if the inoculum is introduced deep into the meat. For this a tall column of meat is essential as anaerobic conditions prevail only where there are particles of meat. There need only be sufficient broth to extend about ½ inch above the meat.)

Malt agar

Dissolve 40 g of malt extract in 1 litre of water and add 20 g of agar.

Completely dissolve these by steaming.

Filter through muslin and adjust the pH to 5·4.

Autoclave at 115°C for 15 minutes.

Sabourard's agar

Add 40 g of glucose to 20 g of agar and 10 g of peptone.

Dissolve these in 1 litre of water in an autoclave.

Filter through muslin and adjust the pH to 5·4.

Autoclave at 115°C for 15 minutes.

Löffler's methylene blue

Saturated solution of methylene blue in alcohol	300 ml
0·01% KOH in water	1,000 ml

Nigrosin

Make up a 10% solution in warm de-ionised water (this may take an hour to dissolve).

Filter the solution.

Add a few drops of 0·5% formalin to act as a preservative.

Gram's iodine

Iodine	1 g
KI	2 g
De-ionised water	300 ml

Crystal violet

Crystal violet	1 g
De-ionised water	99 ml

Carbolfuchsin

Dissolve 5 g of basic fuchsin powder in 25 g of crystalline phenol by placing them in a 1 litre flask over a boiling water bath for 5 minutes. Shake the contents from time to time.

When there is a complete solution, add 50 ml of 95% alcohol and mix thoroughly.

Then add 500 ml of de-ionised water and filter the mixture after allowing it to stand for 24 hours.

Lactophenol containing cotton blue stain

This is a combined fixative, clearing reagent and stain.

Phenol   20 g
De-ionised water   20 ml

Dissolve the phenol completely and add 20 g of lactic acid (= 16·8 ml).

When this has completely dissolved, and 40 g (= 33·3 ml) of pure glycerine.

If the fungus is not coloured, add 20 ml of a 0·05% solution of cotton blue in water.

Store in a brown glass bottle.

This reagent will dissolve varnishes etc, therefore keep it away from the microscope lenses.

Appendix 9

Centrifuges

A centrifuge is an instrument designed to separate materials of different densities from each other by means of centrifugal force. Gravity has the same effect, but a very long time is required to separate materials by gravity alone. A centrifuge allows us to hasten this effect by applying a larger force.

One of the parts to be separated is always a liquid and the other is usually a solid and occasionally a gas.

Centrifugal force

The size of this force depends on the radius of the circle and the speed of rotation.

Let a wheel turn through a circle with an angular velocity of ω radians per second (a radian is the part of the circle circumference equal in length to the radius).

The velocity of a point on the surface is $v = \omega R$, where R represents the radius.

The velocity, expressed in units of length per unit of time, does not change, but the point on the surface is constantly changing direction, this point is subjected to an acceleration of $\omega^2 R$.

283

The angular velocity (ω) can be converted to revolutions per second as 2π radians is one full circle.

The outward centrifugal force is equal and opposite to the centripetal force, i.e. an inward force which accelerates a mass towards the centre of the circle.

Centripetal force, $F_c = m\,\omega^2 R$, where m represents mass, or in terms of revolutions per second, $m(2\pi N)^2 R = m4\pi^2 N^2 R$ where N is the number of revolutions per second.

Centrifugal force is equal to this in size.

Usually the relative centrifugal force is calculated from this.

$$RCF = \frac{F_c}{F_g} = \frac{m4\pi^2 N^2 R}{mg}$$ where g represents the acceleration due to gravity (980 cm/sec^2).

The *RCF* is expressed as so many times g.

All new centrifuges must be calibrated. N can be determined with a stroboscope and R can be measured.

It should be stressed here that increase in dial setting is not proportional to increase in *RCF* value. The calibration graph is always a curve.

Centrifuge heads

The centrifuge may have free swinging heads or fixed angle heads. The movement of the particles down the tube is opposed by the viscosity of the liquid. In a free swinging head the particles must move the whole length of the tube against the viscosity.

The advantage of the angular head is that the time required for centrifugation is considerably reduced as the material does not have to travel down the full length of the tube. The fixed angle tubes are held in a metal head which in its rotation offers little resistance to air and so it reaches a greater speed and is less liable to warm up due to friction.

Disadvantage of the fixed angle head is that a line of deposit is formed on the side of the tube and when the suspending fluid is decanted off it is difficult to prevent the loss of some of the sediment in the fluid. Neither is it possible to use it for quantitative work.

Method of use

1. **The tubes must be placed in the centrifuge head in pairs that have been accurately balanced.** If there are an odd number of tubes a tube of water must be used to balance the odd tube. For school use it is usually satisfactory to merely check that the liquids are the same level in each tube.

2. If using cotton wool plugs, check that they will not be drawn down into the tubes. In a swing out centrifuge, spread the plug out over the top of the tube and secure it with an elastic band. In an angle centrifuge, it is sufficient to splay out the top of the plug. Nevertheless, some cotton wool will become detached and

appear in the centrifugate. Aluminium or stainless steel caps can be used to prevent this, and to keep the tubes sterile.

3. After the tubes are in place, close the lid and check that it is secure. **The lid must not be removed when the centrifuge is running.**

4. When switching on the centrifuge, allow the instrument to gather speed gradually, otherwise the life of the instrument will be curtailed.

Generally speaking 600 G is sufficient to throw down most chemical precipitates which will be used; yeasts and fungi require a centrifuge with a maximum value of 1,000–2,000 G; and bacteria require 2,000–4,000 G.

5. Allow the centrifuge to come to a stop. Never use your hand to brake the rotating head. This will tend to redisperse the centrifugate and of course it could cause serious injuries. Wait until the centrifuge stops before attempting to remove the tubes.

Appendix 10

Colorimetry

Colorimetric estimations, which are commonly made by chemists, are also of considerable value in biological work. Colorimetry depends upon the interaction between light and matter. Most materials will absorb light energy of some wavelengths. Should these wavelengths lie within the visible part of the spectrum, we say the substance is coloured because the eye sees only those parts of the spectrum which are transmitted or reflected. If the light absorbed lay in the infra red or ultraviolet range of the spectrum, we would regard it as colourless material. If most of the visible light is absorbed the material appears to be black.

Instruments suitable for use in schools are those which measure absorption in the visible range. (Such instruments are sufficiently sensitive to detect colour in solutions which appear to be colourless to the naked eye.) Ultraviolet light is strongly absorbed by some biological materials, notably proteins, nucleic acids, vitamins and hormones. The instruments for measurement in the ultraviolet range are very expensive as the optics and sample tubes have to be made of quartz, and a special source of energy has to be provided.

By using a colorimeter, the biologist is able to make very accurate and rapid measurements of the concentration of substances in solution. This can be of particular value in enzyme/substrate concentration work and some reactions can be followed directly on the instrument. After suitable calibration against plated samples, it is possible to use the instrument to measure the numbers of bacteria present in a broth culture—a technique which is invaluable as a time-saver in school project work.

The Instrument

In most instruments there is a polychromatic light source. Light passes from the source, through a slit, to the specimen. Any transmitted light is then measured by means of a photo-electric cell. To increase the accuracy, it is usual to place a coloured filter between the light source and the specimen so that it is the absorption of light within a limited range of wavelengths which is measured.

Filters

These are normally chosen from the Ilford spectrum series, or the Chance glass tri-colour set, or the Ilford gelatine tri-coloured set, the two latter being adequate for school use. The response curves of each set are shown in the diagrams.

The correct filter for any determination is of the colour complementary to that of the prepared sample—that is, of the wavelength at which the maximum light absorption occurs in the sample. In doubtful cases, the choice therefore falls on the filter which, with the meter correctly set at zero on the blank solution produces the highest reading with the prepared sample in the light beam.

Test tubes

If results are to be meaningful, it is essential to use matched tubes of suitable size. The tubes are supplied in boxes of one dozen, each box being marked with a number indicating the grading of the tubes. Each tube carries a ground-glass mark for correct orientation in the sample socket.

The sizes and sample requirements of the available tubes are:

$\frac{5}{8}''$ dia. — 7 ml

$\frac{1}{2}''$ dia. — 4 ml

$\frac{1}{4}''$ dia. — 1·5 ml

The choice of tube size depends not only upon the amount of sample available, but also on its concentration. (To produce readings within a given range, tube size is varied inversely to concentration.)

It must be remembered that a test is valid only if it is carried out in a tube of the same size and grading as that used for the 'blank' solution in zeroing the meter and for the standards with which the sample is to be compared.

Test tubes must be optically clean. Handle them only by the rim to avoid finger marks on the section traversed by the light beam; carefully wipe any overflow of sample from the outside of the tube after filling and make sure that tubes are thoroughly cleaned after use.

Perfect cleanliness of optical glassware is essential for the operation of a colorimeter. Routine cleaning may be done with the usual strong acid mixtures. An effective alternative is to use a special

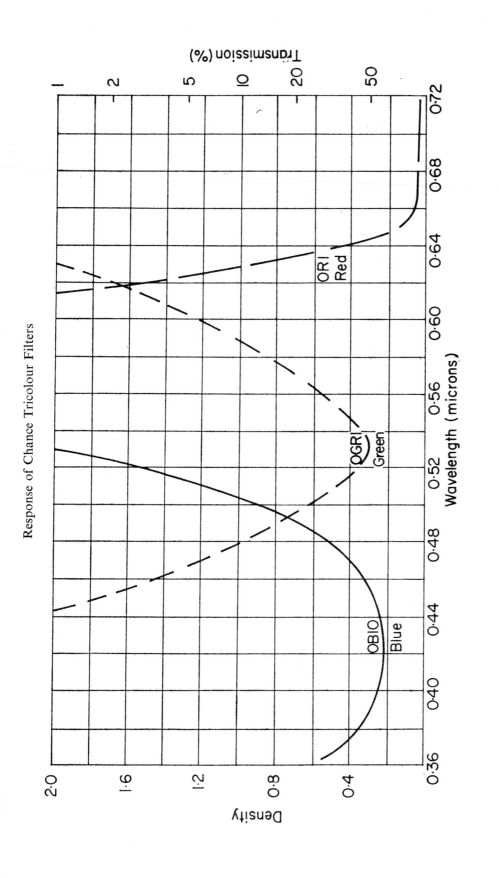

Response of Chance Tricolour Filters

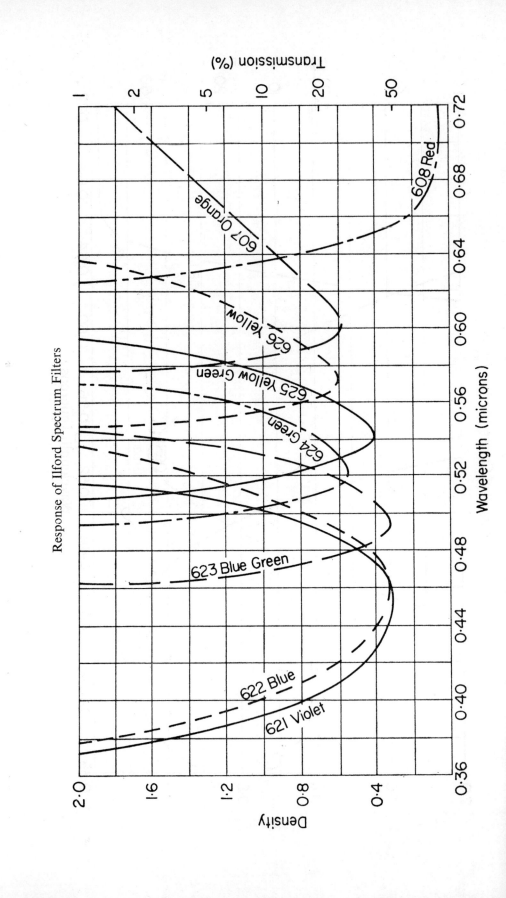

Response of Ilford Spectrum Filters

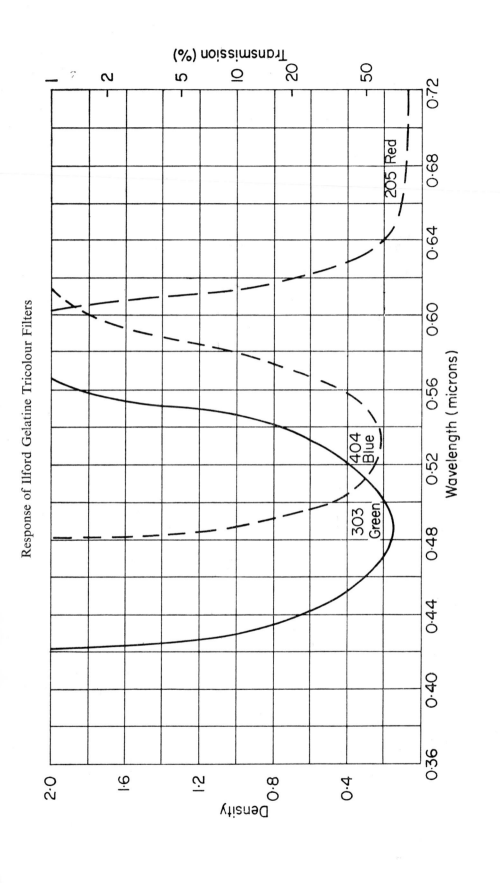

Response of Ilford Gelatine Tricolour Filters

detergent, formulated for laboratory glassware. One such detergent, RBS 25 Concentrate, when used in accordance with the manufacturer's instructions, is suitable. This material is marketed in Great Britain by:

> Medical and Pharmaceutical Departments Ltd,
> 2, Erringham High Road,
> Shoreham-by-Sea,
> Sussex.

Calibration

It is common practice to calibrate the instrument by plotting a series of meter readings against known concentrations. Graphs prepared in this way should carry the following information:

i. The size and grade of the tubes used.

ii The filter number.

iii. The nature of the 'blank' solution used (i.e. whether distilled water or a reagent mixture).

iv. The batch numbers of the reagents used.

Sample preparation

The general principles of sample preparation are well known and universally accepted. Nevertheless, it is stressed that the accuracy of the instrument is of no avail, if it is not matched by the accuracy of the associated techniques. The following points of procedure are therefore listed as being particularly important in the preparation of samples for photo-electric colorimetry.

i. Use analytical grade reagents only.

ii. Ensure that all glassware is clean and dry.

iii. When transferring liquids from vessel to vessel, do so gently to avoid frothing. Air bubbles in the sample will cause inaccuracies.

iv. Measure quantities accurately and avoid wastage during transfer.

'Blank' solutions

The 'blank' solution provides a reference against which the meter may be set to a true zero, as described under the heading 'Method of Use'.

To offset the effect of any slight colour in the reagents themselves, the 'blank' should ideally consist of the mixture of reagents used in preparing the sample. Distilled water may however be used instead, if the reagent mixture is known to be quite colourless.

Method of use

i. With the mains/battery model, check that the 'Mains/battery' switch is in the correct position. Do not connect the instrument to the electrical supply yet.

ii. Switch the 'On/Off' switch on. This permits free swinging of the meter coil, which is short-circuited when the switch is in the 'Off' position.

iii. With the cover closed over the sample aperture to exclude all light from the photocell, check that the meter pointer lies exactly on '∞'. If it does not, adjust the setting as recommended in the makers' instructions.

iv. Set the 'Increase Light' control to minimum, and connect the instrument to the supply. Leave the instrument to stabilise for a few minutes.

v. Select and insert the filter.

vi. Place and orientate a test tube containing the 'blank' solution in the sample socket, and close the cover over it.

vii. Adjust the 'Increase Light' control until the meter reads exactly zero.

viii. Remove the 'blank' solution, insert and orientate the first sample or standard to be determined, and note the meter reading.

ix. Proceed with other determinations in the series, re-checking the zero setting against the 'blank' solution from time to time. The instrument may be left switched on throughout the normal working period, but must in any case be switched off before the instrument is moved.

1. Use analytical grade reagents only.
2. Ensure that all glassware is clean and dry.
3. When transferring liquids from vessel to vessel, do so gently to avoid frothing. Air bubbles in the sample will cause inaccuracies.
4. Measure quantities accurately and avoid wastage during transfer.

Appendix 11

Neon stimulator

This piece of equipment is preferable to the usual induction coil stimulator used in neuromuscular physiology and can be made very cheaply. Coarse adjustment of the discharge frequency is made by switching in different capacitors, fine adjustment being achieved by means of a variable resistance. The intensity of the stimulus delivered to the tissue is controlled by means of a potentiometer.

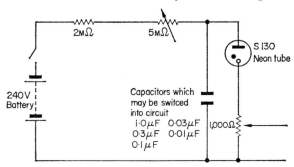

Appendix 12

Varnishing and paper

Kymograph records can be made permanent by the application of a suitable varnish. The usual shellac varnish is slow drying and messy. An excellent substitute is made by dissolving plastic Petri dishes in chloroform to make a 'thin' varnish. The smoked paper may be dipped in this liquid (Shandon dipping tray is ideal), and air dried in one minute. **Carry out the operation in a fume cupboard.**

Appendix 13

Table of chromatographic techniques

Identification of spots

Substances	Solvents
Chlorophyll and carotenoids	100 parts petroleum ether (B.P. 100–120°C), 12 parts acetone. For thin layer techniques, 100 parts of pet. ether (B.P. 60–80°C) to 20 parts of acetone is better.
Plant acids	100 parts of n-butyl formate, 40 parts of 98% formic acid, 10 parts distilled water. Take care with these substances. Add 0·5 g of sodium formate to each 100 ml of solvent and sufficient solid bromophenol blue to turn the solvent a pale orange colour.
Amino acids, sugars and plant acids	28 g of pure crystalline phenol (**use rubber gloves for handling this substance**) should be placed in a separating funnel with 12 ml of distilled water. Add a little NaCl and shake thoroughly. Fill the funnel with coal gas to prevent oxidation and leave for the layers to separate out. This may take an hour. The lower layer is the phenol saturated with water; reject the upper layer.
Amino acids, and sugars	120 parts n-butanol, 30 parts glacial acetic acid, 60 parts distilled water.
Amino acids (2 — way)	For the second solvent use 180 parts ethanol, 10 parts ammonium hydroxide, 10 parts distilled water.
Anthocyanins	40 parts n-butanol, 10 parts glacial acetic acid, 50 parts distilled water. Make up the solvent in a separating funnel and reject the upper layer.

Substances	Methods of extraction from plant material	Development
Chlorophyll and carotenoids	Grind in acetone	No development necessary. Ultraviolet light is useful for viewing the spots. Store the chromatograms in the dark as these pigments are not stable in light.
Plant acids	Grind in 70% ethanol	Dry carefully in a fume cupboard. Sometimes further development is worth while by holding the chromatogram over a bottle of ammonium hydroxide. Do not let the paper get too blue. Sometimes the spots become more clear after a few days.
Amino acids	Grind in water or 80% ethanol	Dry in a fume cupboard. Spray with 2% ninhydrin in acetone. (**Care: ninhydrin is a carcinogen.**) Heat the chromatogram strongly for the spots to appear.
Sugars	Grind in 50% ethanol	Dry the chromatogram in a fume cupboard. Spray with either a 3% solution of para-anisidine hydrochloride in n-butanol plus a few drops of hydrochloric acid or with a 10% solution of resorcinol in acetone plus a few of hydrochloric acid.
Plant acids	Grind in water	Dry the chromatogram in a fume cupboard. Spray with bromothymol blue adjusted to pH 8·5 with sodium hydroxide.
Amino acids	Grind in water or 80% ethanol	Dry carefully as butanol is highly inflammable. Spray with 2% ninhydrin in acetone. Heat strongly for the spots to appear.
Sugars	Grind in 0·1 M sodium acetate in water	Dry carefully as butanol is highly inflammable. Spray with 10% resorcinol in acetone plus a few drops of hydrochloric acid. Heat strongly for the spots to appear.
Amino acids	(as above)	Dry carefully as ethanol is highly inflammable. Spray with 2% ninhydrin in acetone. Heat strongly for the spots to appear.
Anthocyanins	Grind in 80% ethanol	No development is necessary. Treatment with ammonia vapour may intensify the spots.

Appendix 14

Table of R_f values (after Baron, 'Organization in plants', 1967, Arnold)

Identification of spots

Substances	Thin layer chromatography			Paper chromatography		
	Substance	R_f value	Colour	Substance	R_f value	Colour
Chlorophyll and Carotenoids	Chlorophyll b	0·10	Yellow-green	Chlorophyll b	0·45	Green
	Chlorophyll a	0·11	Blue-green	Chlorophyll a	0·65	Blue-green
	Xanthophyll	0·22	Yellow	Xanthophyll	0·71	Yellow-brown
	Phaeophytin	0·28	Grey	Phaeophytin	0·83	Grey
	Carotene	0·90	Yellow	Carotene	0·95	Yellow
	When using petroleum ether (60–80°C)					
	Chlorophyll b	0·17	Yellow-green			
	Xanthophyll (?)	0·19	Yellow			
	Chlorophyll a	0·23	Blue-green			
	Xanthophyll	0·35	Yellow			
	Phaeophytin	0·44	Grey			
	Carotene	0·96	Yellow			
Plant acids	Tartaric acid	0·32	Yellow	Tartaric acid	0·20	Yellow
	Citric acid	0·50	against	Citric acid	0·25	against
	Malic acid	0·61	purple back-	Oxalic acid	0·32	purple back-
	Pyruvic acid	0·85	ground	Malic acid	0·37	ground
	Succinic acid	0·92		Succinic acid	0·57	
Amino acids	Glycine	0·29	Orange-red	Glutamic acid	0·38	Orange-red
	Arginine	0·32	Deep red-purple	Glycine	0·50	Brown-purple
	Cystine	0·32	Pink	Tyrosine	0·66	Deep purple
	Valine	0·41	Red-purple	Arginine	0·70	Red-purple
	Phenyl-alanine	0·52	Brown	Alanine	0·72	Blue-purple
				Leucine	0·91	Deep purple
				Proline	0·95	Yellow
Sugars				Glucose	0·31	Red with
				Sucrose	0·35	resorcinol
				Fructose	0·46	yellow brown with anisidine
Plant acid				Tartaric acid	0·23	Yellow against a
				Citric acid	0·32	blue back-ground
				Oxalic acid	0·35	
				Malic acid	0·43	
				Pyruvic acid	0·59	
				Succinic acid	0·60	
Amino acids	Glycine	0·20	Orange-red			
	Arginine	0·23	Deep red-purple			
	Tyrosine	0·25	Purple			
	Cystine	0·26	Pink			
	Valine	0·39	Red-purple			
Sugars	Sucrose	0·24	Yellow-brown			
	Fructose	0·31	Yellow-brown			
	Glucose	0·43	Yellow-brown			
Anthocyanins				Delphinidin	0·59	Blue-purple
				Pelargonidin	0·73	Bright red
				Peonidin	0·74	Magenta
				Cyanidin	0·79	Mauve-purple

Appendix 15

Photographing chromatograms

Reflex contact document paper should be used. There are two grades, rapid and normal. This may be handled in a bright yellow safe light. Opening the door does not matter, unless the light falls directly on to the paper.

Place the chromatogram in close contact with the photographic paper.

Expose the paper by holding it 1 metre from the lamp. On the average it must be exposed for 8 to 12 seconds for Whatman number 1 filter paper, and 3 minutes for Whatman number 3 filter paper. If the spots are very thick, it may be possible to make several exposures to resolve them. Exposure times are usually not critical. Very thick spots may be made less opaque by wiping with carbon tetrachloride or ethyl acetate.

Develop the film in the following solution:

Metol	3 g
Na_2SO_4 (anhydrous)	50 g
Na_2CO_3 (anhydrous)	70 g
Hydroquinone	13 g
NaBr	1 g
Distilled water	1 litre

30 to 60 seconds is usually a suitable time for development.

Dip the print in water and then in an acid fixing bath made up as follows:

$Na_2S_2O_3$, $5H_2O$	250 g
$K_2S_2O_5$	25 g
Water	1 litre

The print should only be in this bath for a few seconds.

Then wash the print for a few minutes in running water.

Blot it dry and hang up to completely dry. A short wash in acetone may be useful for the rapid drying of the print.

Appendix 16

The mathematical treatment of data

Describing a distribution in mathematical terms

The distributions met with in biology are usually like that shown in diagram. There is a peak in the middle and the distribution falls off on each side more or less symmetrically at higher and lower values of the variable.

295

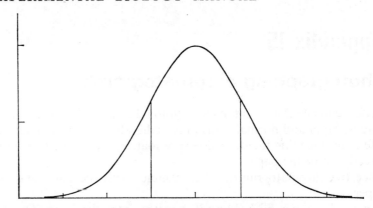

The position of the peak can be described in three ways

(a) **Mode.** This is the most frequent class. This may be misleading as it may be near one of the extremes.

(b) **Median.** This is the middle observation, i.e. if there are 29 observations, the median will be the 15th.

(c) **Mean.** This is the arithmetic mean or average.
To place less stress on the extreme values, arrange all the values in ascending order, and multiply the first by one, the second by four, the third by two, the next by four, the next by two and so on, finally multiply the penultimate score by 4 and the last by one. Add all these results together and divide by the sum of the ones, fours and twos.
There must be an odd number of readings.

The spread of a distribution can be expressed in several ways

(a) **Range.** This is the difference between the highest and the lowest reading. Note that these are often atypical.

(b) **Semi-interquartile range.** This places less emphasis on the extreme values. If a line is drawn through the median of the distribution, it will divide it into two halves of equal area. If lines are then drawn to divide each half into quarters, these lines then become the upper and lower quartile points, i.e. if there are 29 observations, the median will be the 15th and the quartile points will be the 7th and 22nd observations. The interquartile range is then measured from the lower to the upper quartile points.

(c) **Standard deviation.** The individual observations show a deviation about a mean. The scatter is expressed as the standard deviation. This takes all the observations into account.
Standard deviation

$$= \sqrt{\left(\frac{\text{Sum of squares of deviations from mean}}{\text{Number of observations minus one}}\right)}$$

or, $\text{S.D.} = \sqrt{\left(\frac{\Sigma(x - \bar{x})^2}{N - 1}\right)}$

where x represents each observed value taken in turn,

$\bar{x}$ represents the mean value,

N represents the number of observations.

Standard error

If the phenomen being investigated is a very variable one, producing scattered results, no two series of measurements will give exactly the same mean value. Also the more variable the observations, the greater the number of observations which will be required. The reliability of the mean therefore depends on the standard deviation and the number of observations.

$$\text{Standard error} = \frac{\text{standard deviation}}{\sqrt{N}}$$

A mean value is therefore expressed as the mean $\pm$ standard error, e.g. if the mean is quoted as 8 ± 0.4 any determination of the mean will give a result between 7·6 and 8·4.

Samples

The normal distribution curve is based on the total population, but observations in biology can rarely consider this. Usually only a very small part of the population can be observed.

It must therefore be decided whether the results are similar to those that would be obtained if the whole population could be sampled.

Tests of significance

(a) Standard error of difference

Standard error indicates how much the mean of other similar samples drawn from the same population would vary. It reflects the distribution that would be obtained if chance alone operated. In comparing the means of two sets of data, we must know whether there is any real difference between them or not. We must decide whether there is any significance in a slight difference between two means. The difference between two means can be tested for significance by calculating the standard error of difference.

$$\text{S.E.}_{\text{diff}} = \sqrt{\text{Sum of the squares of S.E.'s.}}$$

The difference between the means can then be tested for significance in two ways:

(i) **Large number of observations.** When the difference between the means is more than twice the standard error of difference, the means are significantly different.

For this test to be of any use a minimum number of 10 observations must be made.

(ii) **t-test.** $t = \dfrac{\text{Difference between the means}}{\text{Standard error of the difference}}$

At this point it is necessary to decide on an acceptable level of probability. Absolute certainty can only be achieved if the total population is being observed. For most biological work, a 95% probability is usually acceptable. This means that conclusions made as a result of the analysis will be correct 19 times out of every 20. Refer to the t tables, calculating the degrees of freedom which equal the total number of observations minus one for each set added together. For significance, the t ratio should be greater than the ratio listed in the table opposite the appropriate number of degrees of freedom.

t can also be calculated from the ratio $\dfrac{\text{deviation from mean}}{\text{standard deviation}}$

and $\dfrac{\text{difference from expected ratio}}{\text{standard error}}$

t test can be applied to a very few observations.

Chi-square test

Whereas the t test is used to determine if there is any significance in a deviation from a mean value when a sample is taken from a population, and when the variable has numerical values, chi-squared is used to measure the deviation of ratios. It is commonly used when the results of an experiment are placed in categories and the number of observations in each category are counted.

Chi-square is found by squaring each difference between the number expected and the number observed and then dividing by the expected number in each class, and finally adding the quotients together.

$$\chi^2 = \sum \frac{(\text{Observed} - \text{Expected})^2}{\text{Expected}}$$

The number of degrees of freedom is found by subtracting one from the number of categories.

The probability that the deviation obtained is due to chance alone, can then be determined from the tables reading along the line for the appropriate number of degrees of freedom, e.g. for an experiment with 2 degrees of freedom and a chi-square value of 0·211, there is a probability of 90% that deviations of this size will occur by chance alone. Therefore the hypothesis being applied to produce the expected ratios need not be rejected. Where a value of chi-square is not actually given in the tables, read the probability for the nearest value given in the tables.

The correlation of two variables

Scatter diagrams

This gives a visual impression of the distribution when for example the weight of pigs at birth is being compared with their length. Length is plotted along one axis and weight along the other. This is a convenient way of displaying results but it does not give a reliable way of deciding whether there is a significant correlation.

298

Correlation coefficient

$$r = \frac{\Sigma[(x - \bar{x})(y - \bar{y})]}{\sqrt{[\Sigma(x - \bar{x})^2 \Sigma(y - \bar{y})^2]}}$$

where one variable is represented by x and the other by y, and $\bar{x}$ and $\bar{y}$ are the means of the two variables. $(x - \bar{x})$ and $(y - \bar{y})$ are the deviations from the means of the two variables.

The correlation coefficient can range from $+1$ to -1. Zero means no correlation, $+1$ means a perfect positive correlation, and -1 means a perfect negative correlation. Intermediate values mean varying degrees of correlation.

Significance of the correlation coefficient

This may be measured from correlation coefficient tables.
If the value from the tables is less than the calculated one the latter may be said to show correlation.

Bibliography

*These contain practical details.

Microscopy

*G. W. White, *Introduction to Microscopy*, Butterworths
*V. E. Coslett, *Modern Microscopy*, Bell
*J. D. Casartelli, *Microscopy for Students*, McGraw-Hill
*A. L. E. Barron, *Using the Microscope*, Chapman & Hall
**Phase Contrast Microscopes (Phase Contrast 'conversion kits')*, Education in Science, Nov, 1968

Cytology

S. Hurry, *Microstructure of Cells*, Murray
W. A. Jensen, *The Plant Cell*, Macmillan
E. H. Mercer, *Cells and Cell Structure*, Hutchinson
J. Paul, *Cell Biology*, Heinemann
   The Living Cell-Readings from Scientific American, Freeman
C. P. Swanson, *The Cell*, Prentice-Hall

Microbiology

*B.S.C.S. laboratory block and teachers supplement, *Microbes, their Growth, Nutrition and Interaction*, Harrap
*Seeley and Vandemark, *Microbes in Action: selected exercises*, Freeman
*Collins, *Laboratory Techniques in Microbiology*, Butterworths
*Cruikshank, *Medical Microbiology*, Livingstone
*R. R. Gillies, *Bacteriology Illustrated*, Livingstone
A. J. Salle, *Bacteriology*, McGraw-Hill
W. B. Hugo, *An Introduction to Microbiology*, Harrap
Carpenter, *Microbiology*, Saunders
Ingram, *An Introduction to the Biology of Yeasts*, Pitman
*D. Pramer, *Life in Soil*, Heath
*Dade and Gunnell, *Classwork with Fungi*, Commonwealth Mycological Institute, Kew
K. Vickerman, *The Protozoa*, Murray

BIBLIOGRAPHY

K. M. Jack, S. J. Lapage and J. W. Shewan, *Supply of Bacteriological Cultures to Schools*, Education in Science, Feb. 1969
*D. A. Patterson, *The Continuous Culture of Micro-organisms*, S.S.R. June 1968
Ten Practical Experiments for Use in Classwork, Oxoid, London

Genetics

*G. Haskell, *Practical Heredity with Drosophila*, Oliver & Boyd
*Strickberger, *Experiments in Genetics with Drosophila*, Wiley
*Darlington and La Cour, *The Handling of Chromosomes*, Allen & Unwin
*McLeish and Snoad, *Looking at Chromosomes*, Macmillan
A. M. Winchester, *Heredity, an Introduction to Genetics*, Harrap
B. G. Ashton, *Genes, Chromosomes and Evolution*, Longmans
Bonner and Mills, *Heredity*, Prentice-Hall
Srb and Owen, *General Genetics*, Freeman
H. L. K. Whitehouse, *Towards an Understanding of the Mechanism of Heredity*, Arnold
*B.S.C.S. laboratory block and teachers supplement, *Genetic Continuity*, Harrap
R. J. Berry, *Genetics—Teach Yourself*, E.U.P.
Lawson and Burmester, *Programmed Genetics* (2 volumes), Harrap
*J. G. Thomas, *Dissection of the Locust*, Witherby
J. R. S. Fincham, *Microbial and Molecular Genetics*, E.U.P.
S. A. Cook, *Reproduction, Heredity and Sexuality*, Macmillan
L. H. Penrose, *Human Heredity*, Heinemann
*M. E. Wallace, J. B. Gibson and P. J. Kelly, *Teaching Genetics: The Practical Problems of Breeding Investigations*, Journal of Biological Education Dec. 1968
*Rhoste, *The Use of Tribolium Beetles for Class Practical Work in Genetics*, Journal of Biological Education, Dec. 1968

Biochemistry

*Abbott and Andrews, *Introduction to Chromatography*, Longmans
*Smith and Feinberg, *Paper and Thin Layer Chromatography*, Shandon Scientific Company
E. Baldwin, *Dynamic Aspects of Biochemistry*, Cambridge
E. Baldwin, *Nature of Biochemistry*, Cambridge
E. Baldwin, *An Introduction to Comparative Biochemistry*, Cambridge
E. O'F. Walsh, *Introduction to Biochemistry*, E.U.P.
F. R. Jevons, *The Biochemical Approach to Life*, Allen & Unwin
D. A. Coult, *Molecules and Cells*, Longmans
H. R. Downes, *The Chemistry of Living Cells*, Longmans
J. C. Kendrew, *The Thread of Life*, Bell
K. Harrison, *A Guide Book to Biochemistry*, Cambridge
*Zwarenstein and Van der Schyff, *Practical Biochemistry*, Livingstone
E. H. White, *Chemical Background for the Biological Sciences*, Prentice-Hall
Cell Physiology and Biochemistry, Prentice-Hall

Plant physiology

*W. M. M. Baron, *Organization in Plants*, Arnold
G. A. Strafford, *Essentials of Plant Physiology*, Heinemann
W. M. M. Baron, *Physiological Aspects of Water and Plant Life*, Heinemann
*B.S.C.S. laboratory block, *Regulation in Plants by Hormones*, Harrap
*W. O. James, *An Introduction to Plant Physiology*, Oxford
G. E. Fogg, *Plant Growth and Form*, Pelican
F. C. Steward, *About Plants—Topics in Plant Biology*, Addison-Wesley

*P. A. M. Paice, *Radioisotopes in School Biology*, S.S.R. Sept. 1968
*A. G. Smithers and K. Wilson, *Metabolic Absorption of Mineral Salts*, Journal of Biological Education, Sept. 1968

Animal physiology

Marshall and Hughes, *Physiology of Mammals and other Vertebrates*, Cambridge
Clegg and Clegg, *A Biology of the Mammal*, Heinemann
G. M. Hughes, *A Comparative Physiology of Vertebrate Respiration*, Heinemann
Green, *An Introduction to Human Physiology*, Oxford
Schmidt-Nielsen, *Animal Physiology*, Prentice-Hall
J. B. Jennings, *Feeding, Assimilation and Digestion*, Pergamon
A. P. M. Lockwood, *Animal Body Fluids and their Regulation*, Heinemann
G. Chapman, *Body Fluids and their Movement*, Arnold
Lee and Knowles, *Animal Hormones*, Hutchinson
E. J. W. Barrington, *Hormones and their Evolution*, E.U.P.
L. E. Bayliss, *Living Control Systems*, E.U.P.
A. S. Mason, *Health and Hormones*, Pelican
Whitfield, *An Introduction to Electronics for Physiological Workers*, Macmillan
*R. B. Clark, *Practical Course in Experimental Zoology*, Wiley
*D. Arthur, *Looking at Animals Again*, Freeman

Animal behaviour

A. W. G. Manning, *An Introduction to Animal Behaviour*, Arnold
Dethier and Stellar, *Animal Behaviour*, Prentice-Hall
Psychobiology—readings from Scientific American, Freeman
J. D. Carthy, *Animal Behaviour*, Aldus

Statistics

Bishop, *Statistics for Biology*, Longmans
R. C. Campbell, *Statistics for Biologists*, Cambridge
Bailey, *Statistical Methods in Biology*, E.U.P.
K. Mather, *Statistical Analysis in Biology*, Methuen

Culture methods

*Luty, Welch, Galtorff and Needham, *Culture Methods for Invertebrate Animals* Dover.
*U.F.A.W. Handbook, *Care and Management of Laboratory Animals*, Livingstone
(Section 2 on rodents, lagomorphs, and insectivores is published separately. Section 4 on birds, reptiles, amphibia, fish and insects is also published separately)

General use

*H. A. Peacock, *Elementary Microtechnique*, Arnold
L. J. Hale, *Biological Laboratory Data*, Methuen

International Atomic Weights

The following atomic weights are based on the exact number 12 for the carbon isotope 12, as agreed between the International Unions of Pure and Applied Physics and of Pure and Applied Chemistry.

Name	Symbol	At. No.	At. wt.	Name	Symbol	At. No.	At. wt.
Actinium	Ac	89	—	Mercury	Hg	80	200.59
Aluminium	Al	13	26.9815	Molybdenum	Mo	42	95.94
Americium	Am	95	—	Neodymium	Nd	60	144.24
Antimony	Sb	51	121.75	Neon	Ne	10	20.183
Argon	Ar	18	39.948	Neptunium	Np	93	—
Arsenic	As	33	74.9216	Nickel	Ni	28	58.71
Astatine	At	85	—	Niobium	Nb	41	92.906
Barium	Ba	56	137.34	Nitrogen	N	7	14.0067
Berkelium	Bk	97	—	Nobelium	No	102	—
Beryllium	Be	4	9.0122	Osmium	Os	76	190.2
Bismuth	Bi	83	208.980	Oxygen	O	8	15.9994 ± 0.0001 a
Boron	B	5	10.811 ± 0.003 a	Palladium	Pd	46	106.4
Bromine	Br	35	79.909 b	Phosphorus	P	15	30.9738
Cadmium	Cd	48	112.40	Platinum	Pt	78	195.09
Caesium	Cs	55	135.905	Plutonium	Pu	94	—
Calcium	Ca	20	40.08	Polonium	Po	84	—
Californium	Cf	98	—	Potassium	K	19	39.102
Carbon	C	6	12.01115 ± 0.00005 a	Praseodymium	Pr	59	140.907
Cerium	Ce	58	140.12	Promethium	Pm	61	—
Chlorine	Cl	17	35.453 b	Protoactinium	Pa	91	—
Chromium	Cr	24	51.996 b	Radium	Ra	88	—
Cobalt	Co	27	58.9332	Radon	Rn	86	—
Copper	Cu	29	63.54	Rhenium	Re	75	186.2
Curium	Cm	96	—	Rhodium	Rh	45	102.905
Dysprosium	Dy	66	162.50	Rubidium	Rb	37	85.47
Einsteinium	Es	99	—	Ruthenium	Ru	44	101.07
Erbium	Er	68	167.26	Samarium	Sm	62	150.35
Europium	Eu	63	151.96	Scandium	Sc	21	44.956
Fermium	Fm	100	—	Selenium	Se	34	78.96
Fluorine	F	9	18.9984	Silicon	Si	14	28.086 ± 0.001 a
Francium	Fr	87	—	Silver	Ag	47	107.870 b
Gadolinium	Gd	64	157.25	Sodium	Na	11	22.9898
Gallium	Ga	31	69.72	Strontium	Sr	38	87.62
Germanium	Ge	32	72.59	Sulphur	S	16	32.064 ± 0.003 a
Gold	Au	79	196.967	Tantalum	Ta	73	180.948
Hafnium	Hf	72	178.49	Technetium	Tc	43	—
Helium	He	2	4.0026	Tellurium	Te	52	127.60
Holmium	Ho	67	164.930	Terbium	Tb	65	158.924
Hydrogen	H	1	1.00797 ± 0.00001 a	Thallium	Tl	81	204.37
Indium	In	49	114.82	Thorium	Th	90	232.038
Iodine	I	53	126.9044	Thulium	Tm	69	168.934
Iridium	Ir	77	192.2	Tin	Sn	50	118.69
Iron	Fe	26	55.847 b	Titanium	Ti	22	47.90
Krypton	Kr	36	83.80	Tungsten	W	74	183.85
Lanthanum	La	57	138.91	Uranium	U	92	238.03
Lead	Pb	82	207.19	Vanadium	V	23	50.942
Lithium	Li	3	6.939	Xenon	Xe	54	131.30
Lutetium	Lu	71	174.97	Ytterbium	Yb	70	173.04
Magnesium	Mg	12	24.312	Yttrium	Y	39	88.905
Manganese	Mn	25	54.9380	Zinc	Zn	30	65.37
Mendelevium	Md	101	—	Zirconium	Zr	40	91.22

a Atomic weights so designated are known to be variable because of natural variations in isotopi composition. The observed ranges are:

Hydrogen	±0.00001	Oxygen	±0.0001
Boron	±0.003	Silicon	±0.001
Carbon	±0.00005	Sulphur	±0.003

b Atomic weights so designated are believed to have the following experimental uncertainties:

Chlorine	±0.001	Bromine	±0.002
Chromium	±0.001	Silver	±0.003
Iron	±0.003		

For other elements the last digit is believed to be reliable to ±0.5.

[Table reproduced from I.U.P.A.C. Bulletin number 14B, with permission from Butterwort Scientific Publications, Publishers to the International Union of Pure and Applied Chemistry.]

HEINEMANN'S STATISTICAL TABLES

χ^2

Percentage points of the χ^2 distribution

ν \ Q	0·1	0·5	1·0	2·5	5·0	10·0	25·0	50·0	75·0	90·0	95·0	97·5	99·0	99·5
1	10·828	7·879	6·635	5·024	3·841	2·706	1·323	0·4549	0·1015	0·01579	3932.10^{-6}	9821.10^{-7}	1571.10^{-7}	3927.10^{-8}
2	13·816	10·597	9·210	7·378	5·991	4·605	2·773	1·386	0·5754	0·2107	0·1026	0·05064	0·02010	0·01003
3	16·266	12·838	11·345	9·348	7·815	6·251	4·108	2·366	1·213	0·5844	0·3518	0·2158	0·1148	0·07172
4	18·467	14·860	13·277	11·143	9·488	7·779	5·385	3·357	1·923	1·064	0·7107	0·4844	0·2971	0·2070
5	20·515	16·750	15·086	12·833	11·070	9·236	6·626	4·351	2·675	1·610	1·145	0·8312	0·5543	0·4117
6	22·458	18·548	16·812	14·449	12·592	10·645	7·841	5·348	3·455	2·204	1·635	1·237	0·8721	0·6757
7	24·322	20·278	18·475	16·013	14·067	12·017	9·037	6·346	4·255	2·833	2·167	1·690	1·239	0·9893
8	26·125	21·955	20·090	17·535	15·507	13·362	10·219	7·344	5·071	3·490	2·733	2·180	1·646	1·344
9	27·877	23·589	21·666	19·023	16·919	14·684	11·389	8·343	5·899	4·168	3·325	2·700	2·088	1·735
10	29·588	25·188	23·209	20·483	18·307	15·987	12·549	9·342	6·737	4·865	3·940	3·247	2·558	2·156
11	31·264	26·757	24·725	21·920	19·675	17·275	13·701	10·341	7·584	5·578	4·575	3·816	3·053	2·603
12	32·909	28·300	26·217	23·337	21·026	18·549	14·845	11·340	8·438	6·304	5·226	4·404	3·571	3·074
13	34·528	29·819	27·688	24·736	22·362	19·812	15·984	12·340	9·299	7·041	5·892	5·009	4·107	3·565
14	36·123	31·319	29·141	26·119	23·685	21·064	17·117	13·339	10·165	7·790	6·571	5·629	4·660	4·075
15	37·697	32·801	30·578	27·488	24·996	22·307	18·245	14·339	11·036	8·547	7·261	6·262	5·229	4·601
16	39·252	34·267	32·000	28·845	26·296	23·542	19·369	15·338	11·912	9·312	7·962	6·908	5·812	5·142
17	40·790	35·718	33·409	30·191	27·587	24·769	20·489	16·338	12·792	10·085	8·672	7·564	6·408	5·697
18	42·312	37·156	34·805	31·526	28·869	25·989	21·605	17·338	13·675	10·865	9·390	8·231	7·015	6·265
19	43·820	38·582	36·191	32·852	30·143	27·204	22·718	18·338	14·562	11·651	10·117	8·907	7·633	6·844
20	45·315	39·997	37·566	34·170	31·410	28·412	23·828	19·337	15·452	12·443	10·851	9·591	8·260	7·434
21	46·797	41·401	38·932	35·479	32·670	29·615	24·935	20·337	16·344	13·240	11·591	10·283	8·897	8·034
22	48·268	42·796	40·289	36·781	33·924	30·813	26·039	21·337	17·240	14·041	12·338	10·982	9·542	8·643
23	49·728	44·181	41·638	38·076	35·172	32·007	27·141	22·337	18·137	14·848	13·090	11·688	10·196	9·260
24	51·179	45·558	42·980	39·364	36·415	33·196	28·241	23·337	19·037	15·659	13·848	12·401	10·856	9·886
25	52·620	46·928	44·314	40·646	37·652	34·382	29·339	24·337	19·939	16·473	14·611	13·120	11·524	10·520
26	54·052	48·290	45·642	41·923	38·885	35·563	30·434	25·336	20·843	17·292	15·379	13·844	12·198	11·160
27	55·476	49·645	46·963	43·194	40·113	36·741	31·528	26·336	21·749	18·114	16·151	14·573	12·879	11·808
28	56·892	50·993	48·278	44·461	41·337	37·916	32·620	27·336	22·657	18·939	16·928	15·308	13·565	12·461
29	58·302	52·336	49·588	45·722	42·557	39·087	33·711	28·336	23·567	19·768	17·708	16·047	14·256	13·121
30	59·703	53·672	50·892	46·979	43·773	40·256	34·800	29·336	24·478	20·599	18·493	16·791	14·954	13·787
40	73·402	66·766	63·691	59·342	55·758	51·805	45·616	39·335	33·660	29·050	26·509	24·433	22·164	20·707
50	86·661	79·490	76·154	71·420	67·505	63·167	56·334	49·335	42·942	37·689	34·764	32·357	29·707	27·991
60	99·607	91·952	88·379	83·298	79·082	74·397	66·981	59·335	52·294	46·459	43·188	40·482	37·485	35·535
70	112·317	104·215	100·425	95·023	90·531	85·527	77·577	69·334	61·698	55·329	51·739	48·758	45·442	43·275
80	124·839	116·321	112·329	106·629	101·879	96·578	88·130	79·334	71·144	64·278	60·391	57·153	53·540	51·172
90	137·208	128·299	124·116	118·136	113·145	107·565	98·650	89·334	80·625	73·291	69·126	65·647	61·754	59·196
100	149·449	140·169	135·807	129·561	124·342	118·498	109·141	99·334	90·133	82·358	77·929	74·222	70·065	67·328

$$Q = 100 \cdot \frac{1}{\Gamma\left(\frac{\nu}{2}\right) 2^{\nu/2}} \int_{\chi^2}^{\infty} x^{\frac{1}{2}(\nu-2)} e^{-\frac{1}{2}x} \, dx$$

[For large ν refer $(\sqrt{2\chi^2} - \sqrt{2\nu - 1})$ to tables of normal distribution]

Reproduced with permission from Biometrika Tables for Statisticians

HEINEMANN'S STATISTICAL TABLES

t.

r.

Percentage points of the distribution of the correlation coefficient, r, (when $\rho = 0$).

2Q →	10	5	2	1	0·5	0·1
ν \ Q →	5	2·5	1	0·5	0·25	0·05
ν = 1	·9877	·9969	·9995	·9999	*	*
2	·9000	·9500	·9800	·9900	·9950	·9990
3	·805	·878	·9343	·9587	·9740	·9911
4	·729	·811	·882	·9172	·9417	·9741
5	·669	·754	·833	·875	·9056	·9509
6	·621	·707	·789	·834	·870	·9249
7	·582	·666	·750	·798	·836	·898
8	·549	·632	·715	·765	·805	·872
9	·521	·602	·685	·735	·776	·847
10	·497	·576	·658	·708	·750	·823
11	·476	·553	·634	·684	·726	·801
12	·457	·532	·612	·661	·703	·780
13	·441	·514	·592	·641	·683	·760
14	·426	·497	·574	·623	·664	·742
15	·412	·482	·558	·606	·647	·725
16	·400	·468	·543	·590	·631	·708
17	·389	·456	·529	·575	·616	·693
18	·378	·444	·516	·561	·602	·679
19	·369	·433	·503	·549	·589	·665
20	·360	·423	·492	·537	·576	·652
25	·323	·381	·445	·487	·524	·597
30	·296	·349	·409	·449	·484	·554
35	·275	·325	·381	·418	·452	·519
40	·257	·304	·358	·393	·425	·490
45	·243	·288	·338	·372	·403	·465
50	·231	·273	·322	·354	·384	·443
60	·211	·250	·295	·325	·352	·408
70	·195	·232	·274	·302	·327	·380
80	·183	·217	·257	·283	·307	·357
90	·173	·205	·242	·267	·290	·338
100	·164	·195	·230	·254	·276	·321

* Greater than ·9999

If r calculated from n pairs of observations then $\nu = n - 2$.

Percentage points of the t-distribution

2Q →	30	20	10	5	2	1	0·2	0·1
Q →	15	10	5	2·5	1	0·5	0·1	·05
ν = 1	1·963	3·078	6·314	12·706	31·821	63·657	318·31	636·62
2	1·386	1·886	2·920	4·303	6·965	9·925	22·326	31·598
3	1·250	1·638	2·353	3·182	4·541	5·841	10·213	12·924
4	1·190	1·533	2·132	2·776	3·747	4·604	7·173	8·610
5	1·156	1·476	2·015	2·571	3·365	4·032	5·893	6·869
6	1·134	1·440	1·943	2·447	3·143	3·707	5·208	5·959
7	1·119	1·415	1·895	2·365	2·998	3·499	4·785	5·408
8	1·108	1·397	1·860	2·306	2·896	3·355	4·501	5·041
9	1·100	1·383	1·833	2·262	2·821	3·250	4·297	4·781
10	1·093	1·372	1·812	2·228	2·764	3·169	4·144	4·587
11	1·088	1·363	1·796	2·201	2·718	3·106	4·025	4·437
12	1·083	1·356	1·782	2·179	2·681	3·055	3·930	4·318
13	1·079	1·350	1·771	2·160	2·650	3·012	3·852	4·221
14	1·076	1·345	1·761	2·145	2·624	2·977	3·787	4·140
15	1·074	1·341	1·753	2·131	2·602	2·947	3·733	4·073
16	1·071	1·337	1·746	2·120	2·583	2·921	3·686	4·015
17	1·069	1·333	1·740	2·110	2·567	2·898	3·646	3·965
18	1·067	1·330	1·734	2·101	2·552	2·878	3·610	3·922
19	1·066	1·328	1·729	2·093	2·539	2·861	3·579	3·883
20	1·064	1·325	1·725	2·086	2·528	2·845	3·552	3·850
21	1·063	1·323	1·721	2·080	2·518	2·831	3·527	3·819
22	1·061	1·321	1·717	2·074	2·508	2·819	3·505	3·792
23	1·060	1·319	1·714	2·069	2·500	2·807	3·485	3·767
24	1·059	1·318	1·711	2·064	2·492	2·797	3·467	3·745
25	1·058	1·316	1·708	2·060	2·485	2·787	3·450	3·725
26	1·058	1·315	1·706	2·056	2·479	2·779	3·435	3·707
27	1·057	1·314	1·703	2·052	2·473	2·771	3·421	3·690
28	1·056	1·313	1·701	2·048	2·467	2·763	3·408	3·674
29	1·055	1·311	1·699	2·045	2·462	2·756	3·396	3·659
30	1·055	1·310	1·697	2·042	2·457	2·750	3·385	3·646
40	1·050	1·303	1·684	2·021	2·423	2·704	3·307	3·551
60	1·046	1·296	1·671	2·000	2·390	2·660	3·232	3·460
120	1·041	1·289	1·658	1·980	2·358	2·617	3·160	3·373
∞	1·0364	1·2815	1·6449	1·9600	2·3263	2·5758	3·0902	3·2905

Degrees of Freedom ν

For ν > 30 interpolate using $120/\nu$
Use 2Q for two tailed test

Reproduced with permission from Biometrika Tables for Statisticians

LOGARITHMS

	0	1	2	3	4	5	6	7	8	9	1	2	3	4	5	6	7	8	9
10	0000	0043	0086	0128	0170	0212	0253	0294	0334	0374	4	8	12	17	21	25	29	33	37
11	0414	0453	0492	0531	0569	0607	0645	0682	0719	0755	4	8	11	15	19	23	26	30	34
12	0792	0828	0864	0899	0934	0969	1004	1038	1072	1106	3	7	10	14	17	21	24	28	31
13	1139	1173	1206	1239	1271	1303	1335	1367	1399	1430	3	6	10	13	16	19	23	26	29
14	1461	1492	1523	1553	1584	1614	1644	1673	1703	1732	3	6	9	12	15	18	21	24	27
15	1761	1790	1818	1847	1875	1903	1931	1959	1987	2014	3	6	8	11	14	17	20	22	25
16	2041	2068	2095	2122	2148	2175	2201	2227	2253	2279	3	5	8	11	13	16	18	21	24
17	2304	2330	2355	2380	2405	2430	2455	2480	2504	2529	2	5	7	10	12	15	17	20	22
18	2553	2577	2601	2625	2648	2672	2695	2718	2742	2765	2	5	7	9	12	14	16	19	21
19	2788	2810	2833	2856	2878	2900	2923	2945	2967	2989	2	4	7	9	11	13	16	18	20
20	3010	3032	3054	3075	3096	3118	3139	3160	3181	3201	2	4	6	8	11	13	15	17	19
21	3222	3243	3263	3284	3304	3324	3345	3365	3385	3404	2	4	6	8	10	12	14	16	18
22	3424	3444	3464	3483	3502	3522	3541	3560	3579	3598	2	4	6	8	10	12	14	15	17
23	3617	3636	3655	3674	3692	3711	3729	3747	3766	3784	2	4	6	7	9	11	13	15	17
24	3802	3820	3838	3856	3874	3892	3909	3927	3945	3962	2	4	5	7	9	11	12	14	16
25	3979	3997	4014	4031	4048	4065	4082	4099	4116	4133	2	3	5	7	9	10	12	14	15
26	4150	4166	4183	4200	4216	4232	4249	4265	4281	4298	2	3	5	7	8	10	11	13	15
27	4314	4330	4346	4362	4378	4393	4409	4425	4440	4456	2	3	5	6	8	9	11	13	14
28	4472	4487	4502	4518	4533	4548	4564	4579	4594	4609	2	3	5	6	8	9	11	12	14
29	4624	4639	4654	4669	4683	4698	4713	4728	4742	4757	1	3	4	6	7	9	10	12	13
30	4771	4786	4800	4814	4829	4843	4857	4871	4886	4900	1	3	4	6	7	9	10	11	13
31	4914	4928	4942	4955	4969	4983	4997	5011	5024	5038	1	3	4	6	7	8	10	11	12
32	5051	5065	5079	5092	5105	5119	5132	5145	5159	5172	1	3	4	5	7	8	9	11	12
33	5185	5198	5211	5224	5237	5250	5263	5276	5289	5302	1	3	4	5	6	8	9	10	12
34	5315	5328	5340	5353	5366	5378	5391	5403	5416	5428	1	3	4	5	6	8	9	10	11
35	5441	5453	5465	5478	5490	5502	5514	5527	5539	5551	1	2	4	5	6	7	9	10	11
36	5563	5575	5587	5599	5611	5623	5635	5647	5658	5670	1	2	4	5	6	7	8	10	11
37	5682	5694	5705	5717	5729	5740	5752	5763	5775	5786	1	2	3	5	6	7	8	9	10
38	5798	5809	5821	5832	5843	5855	5866	5877	5888	5899	1	2	3	5	6	7	8	9	10
39	5911	5922	5933	5944	5955	5966	5977	5988	5999	6010	1	2	3	4	5	7	8	9	10
40	6021	6031	6042	6053	6064	6075	6085	6096	6107	6117	1	2	3	4	5	6	8	9	10
41	6128	6138	6149	6160	6170	6180	6191	6201	6212	6222	1	2	3	4	5	6	7	8	9
42	6232	6243	6253	6263	6274	6284	6294	6304	6314	6325	1	2	3	4	5	6	7	8	9
43	6335	6345	6355	6365	6375	6385	6395	6405	6415	6425	1	2	3	4	5	6	7	8	9
44	6435	6444	6454	6464	6474	6484	6493	6503	6513	6522	1	2	3	4	5	6	7	8	9
45	6532	6542	6551	6561	6571	6580	6590	6599	6609	6618	1	2	3	4	5	6	7	8	9
46	6628	6637	6646	6656	6665	6675	6684	6693	6702	6712	1	2	3	4	5	6	7	7	8
47	6721	6730	6739	6749	6758	6767	6776	6785	6794	6803	1	2	3	4	5	5	6	7	8
48	6812	6821	6830	6839	6848	6857	6866	6875	6884	6893	1	2	3	4	4	5	6	7	8
49	6902	6911	6920	6928	6937	6946	6955	6964	6972	6981	1	2	3	4	4	5	6	7	8
50	6990	6998	7007	7016	7024	7033	7042	7050	7059	7067	1	2	3	3	4	5	6	7	8
51	7076	7084	7093	7101	7110	7118	7126	7135	7143	7152	1	2	3	3	4	5	6	7	8
52	7160	7168	7177	7185	7193	7202	7210	7218	7226	7235	1	2	2	3	4	5	6	7	7
53	7243	7251	7259	7267	7275	7284	7292	7300	7308	7316	1	2	2	3	4	5	6	6	7
54	7324	7332	7340	7348	7356	7364	7372	7380	7388	7396	1	2	2	3	4	5	6	6	7

LOGARITHMS

	0	1	2	3	4	5	6	7	8	9	1	2	3	4	5	6	7	8	9
55	7404	7412	7419	7427	7435	7443	7451	7459	7466	7474	1	2	2	3	4	5	5	6	7
56	7482	7490	7497	7505	7513	7520	7528	7536	7543	7551	1	2	2	3	4	5	5	6	7
57	7559	7566	7574	7582	7589	7597	7604	7612	7619	7627	1	2	2	3	4	5	5	6	7
58	7634	7642	7649	7657	7664	7672	7679	7686	7694	7701	1	1	2	3	4	5	5	6	7
59	7709	7716	7723	7731	7738	7745	7752	7760	7767	7774	1	1	2	3	4	5	5	6	7
60	7782	7789	7796	7803	7810	7818	7825	7832	7839	7846	1	1	2	3	4	4	5	6	6
61	7853	7860	7868	7875	7882	7889	7896	7903	7910	7917	1	1	2	3	4	4	5	6	6
62	7924	7931	7938	7945	7952	7959	7966	7973	7980	7987	1	1	2	3	4	4	5	6	6
63	7993	8000	8007	8014	8021	8028	8035	8041	8048	8055	1	1	2	3	3	4	5	5	6
64	8062	8069	8075	8082	8089	8096	8102	8109	8116	8122	1	1	2	3	3	4	5	5	6
65	8129	8136	8142	8149	8156	8162	8169	8176	8182	8189	1	1	2	3	3	4	5	5	6
66	8195	8202	8209	8215	8222	8228	8235	8241	8248	8254	1	1	2	3	3	4	5	5	6
67	8261	8267	8274	8280	8287	8293	8299	8306	8312	8319	1	1	2	3	3	4	5	5	6
68	8325	8331	8338	8344	8351	8357	8363	8370	8376	8382	1	1	2	3	3	4	4	5	6
69	8388	8395	8401	8407	8414	8420	8426	8432	8439	8445	1	1	2	2	3	4	4	5	6
70	8451	8457	8463	8470	8476	8482	8488	8494	8500	8506	1	1	2	2	3	4	4	5	5
71	8513	8519	8525	8531	8537	8543	8549	8555	8561	8567	1	1	2	2	3	4	4	5	5
72	8573	8579	8585	8591	8597	8603	8609	8615	8621	8627	1	1	2	2	3	4	4	5	5
73	8633	8639	8645	8651	8657	8663	8669	8675	8681	8686	1	1	2	2	3	4	4	5	5
74	8692	8698	8704	8710	8716	8722	8727	8733	8739	8745	1	1	2	2	3	4	4	5	5
75	8751	8756	8762	8768	8774	8779	8785	8791	8797	8802	1	1	2	2	3	3	4	5	5
76	8808	8814	8820	8825	8831	8837	8842	8848	8854	8859	1	1	2	2	3	3	4	5	5
77	8865	8871	8876	8882	8887	8893	8899	8904	8910	8915	1	1	2	2	3	3	4	4	5
78	8921	8927	8932	8938	8943	8949	8954	8960	8965	8971	1	1	2	2	3	3	4	4	5
79	8976	8982	8987	8993	8998	9004	9009	9015	9020	9025	1	1	2	2	3	3	4	4	5
80	9031	9036	9042	9047	9053	9058	9063	9069	9074	9079	1	1	2	2	3	3	4	4	5
81	9085	9090	9096	9101	9106	9112	9117	9122	9128	9133	1	1	2	2	3	3	4	4	5
82	9138	9143	9149	9154	9159	9165	9170	9175	9180	9186	1	1	2	2	3	3	4	4	5
83	9191	9196	9201	9206	9212	9217	9222	9227	9232	9238	1	1	2	2	3	3	4	4	5
84	9243	9248	9253	9258	9263	9269	9274	9279	9284	9289	1	1	2	2	3	3	4	4	5
85	9294	9299	9304	9309	9315	9320	9325	9330	9335	9340	1	1	2	2	3	3	4	4	5
86	9345	9350	9355	9360	9365	9370	9375	9380	9385	9390	1	1	2	2	3	3	4	4	5
87	9395	9400	9405	9410	9415	9420	9425	9430	9435	9440	0	1	1	2	2	3	3	4	4
88	9445	9450	9455	9460	9465	9469	9474	9479	9484	9489	0	1	1	2	2	3	3	4	4
89	9494	9499	9504	9509	9513	9518	9523	9528	9533	9538	0	1	1	2	2	3	3	4	4
90	9542	9547	9552	9557	9562	9566	9571	9576	9581	9586	0	1	1	2	2	3	3	4	4
91	9590	9595	9600	9605	9609	9614	9619	9624	9628	9633	0	1	1	2	2	3	3	4	4
92	9638	9643	9647	9652	9657	9661	9666	9671	9675	9680	0	1	1	2	2	3	3	4	4
93	9685	9689	9694	9699	9703	9708	9713	9717	9722	9727	0	1	1	2	2	3	3	4	4
94	9731	9736	9741	9745	9750	9754	9759	9763	9768	9773	0	1	1	2	2	3	3	4	4
95	9777	9782	9786	9791	9795	9800	9805	9809	9814	9818	0	1	1	2	2	3	3	4	4
96	9823	9827	9832	9836	9841	9845	9850	9854	9859	9863	0	1	1	2	2	3	3	4	4
97	9868	9872	9877	9881	9886	9890	9894	9899	9903	9908	0	1	1	2	2	3	3	3	4
98	9912	9917	9921	9926	9930	9934	9939	9943	9948	9952	0	1	1	2	2	3	3	3	4
99	9956	9961	9965	9969	9974	9978	9983	9987	9991	9996	0	1	1	2	2	3	3	3	4

	0	1	2	3	4	5	6	7	8	9	1	2	3	4	5	6	7	8	9
.50	3162	3170	3177	3184	3192	3199	3206	3214	3221	3228	1	1	2	3	4	4	5	6	7
.51	3236	3243	3251	3258	3266	3273	3281	3289	3296	3304	1	2	2	3	4	5	5	6	7
.52	3311	3319	3327	3334	3342	3350	3357	3365	3373	3381	1	2	2	3	4	5	5	6	7
.53	3388	3396	3404	3412	3420	3428	3436	3443	3451	3459	1	2	2	3	4	5	6	6	7
.54	3467	3475	3483	3491	3499	3508	3516	3524	3532	3540	1	2	2	3	4	5	6	6	7
.55	3548	3556	3565	3573	3581	3589	3597	3606	3614	3622	1	2	2	3	4	5	6	7	7
.56	3631	3639	3648	3656	3664	3673	3681	3690	3698	3707	1	2	3	3	4	5	6	7	8
.57	3715	3724	3733	3741	3750	3758	3767	3776	3784	3793	1	2	3	3	4	5	6	7	8
.58	3802	3811	3819	3828	3837	3846	3855	3864	3873	3882	1	2	3	4	4	5	6	7	8
.59	3890	3899	3908	3917	3926	3936	3945	3954	3963	3972	1	2	3	4	5	5	6	7	8
.60	3981	3990	3999	4009	4018	4027	4036	4046	4055	4064	1	2	3	4	5	5	6	7	8
.61	4074	4083	4093	4102	4111	4121	4130	4140	4150	4159	1	2	3	4	5	6	7	8	9
.62	4169	4178	4188	4198	4207	4217	4227	4236	4246	4256	1	2	3	4	5	6	7	8	9
.63	4266	4276	4285	4295	4305	4315	4325	4335	4345	4355	1	2	3	4	5	6	7	8	9
.64	4365	4375	4385	4395	4406	4416	4426	4436	4446	4457	1	2	3	4	5	6	7	8	9
.65	4467	4477	4487	4498	4508	4519	4529	4539	4550	4560	1	2	3	4	5	6	7	8	9
.66	4571	4581	4592	4603	4613	4624	4634	4645	4656	4667	1	2	3	4	5	6	7	9	10
.67	4677	4688	4699	4710	4721	4732	4742	4753	4764	4775	1	2	3	4	5	7	8	9	10
.68	4786	4797	4808	4819	4831	4842	4853	4864	4875	4887	1	2	3	4	6	7	8	9	10
.69	4898	4909	4920	4932	4943	4955	4966	4977	4989	5000	1	2	3	5	6	7	8	9	10
.70	5012	5023	5035	5047	5058	5070	5082	5093	5105	5117	1	2	4	5	6	7	8	9	11
.71	5129	5140	5152	5164	5176	5188	5200	5212	5224	5236	1	2	4	5	6	7	8	10	11
.72	5248	5260	5272	5284	5297	5309	5321	5333	5346	5358	1	2	4	5	6	7	9	10	11
.73	5370	5383	5395	5408	5420	5433	5445	5458	5470	5483	1	3	4	5	6	8	9	10	11
.74	5495	5508	5521	5534	5546	5559	5572	5585	5598	5610	1	3	4	5	6	8	9	10	12
.75	5623	5636	5649	5662	5675	5689	5702	5715	5728	5741	1	3	4	5	7	8	9	10	12
.76	5754	5768	5781	5794	5808	5821	5834	5848	5861	5875	1	3	4	5	7	8	9	11	12
.77	5888	5902	5916	5929	5943	5957	5970	5984	5998	6012	1	3	4	5	7	8	10	11	12
.78	6026	6039	6053	6067	6081	6095	6109	6124	6138	6152	1	3	4	6	7	8	10	11	13
.79	6166	6180	6194	6209	6223	6237	6252	6266	6281	6295	1	3	4	6	7	9	10	11	13
.80	6310	6324	6339	6353	6368	6383	6397	6412	6427	6442	1	3	4	6	7	9	10	12	13
.81	6457	6471	6486	6501	6516	6531	6546	6561	6577	6592	2	3	5	6	8	9	11	12	14
.82	6607	6622	6637	6653	6668	6683	6699	6714	6730	6745	2	3	5	6	8	9	11	12	14
.83	6761	6776	6792	6808	6823	6839	6855	6871	6887	6902	2	3	5	6	8	9	11	13	14
.84	6918	6934	6950	6966	6982	6998	7015	7031	7047	7063	2	3	5	6	8	10	11	13	15
.85	7079	7096	7112	7129	7145	7161	7178	7194	7211	7228	2	3	5	7	8	10	12	13	15
.86	7244	7261	7278	7295	7311	7328	7345	7362	7379	7396	2	3	5	7	8	10	12	13	15
.87	7413	7430	7447	7464	7482	7499	7516	7534	7551	7568	2	3	5	7	9	10	12	14	16
.88	7586	7603	7621	7638	7656	7674	7691	7709	7727	7745	2	4	5	7	9	11	12	14	16
.89	7762	7780	7798	7816	7834	7852	7870	7889	7907	7925	2	4	5	7	9	11	13	14	16
.90	7943	7962	7980	7998	8017	8035	8054	8072	8091	8110	2	4	6	7	9	11	13	15	17
.91	8128	8147	8166	8185	8204	8222	8241	8260	8279	8299	2	4	6	7	9	11	13	15	17
.92	8318	8337	8356	8375	8395	8414	8433	8453	8472	8492	2	4	6	8	10	11	13	15	17
.93	8511	8531	8551	8570	8590	8610	8630	8650	8670	8690	2	4	6	8	10	12	14	16	18
.94	8710	8730	8750	8770	8790	8810	8831	8851	8872	8892	2	4	6	8	10	12	14	16	18
.95	8913	8933	8954	8974	8995	9016	9036	9057	9073	9099	2	4	6	8	10	12	14	16	18
.96	9120	9141	9162	9183	9204	9226	9247	9268	9290	9311	2	4	6	8	11	13	15	17	19
.97	9333	9354	9376	9397	9419	9441	9462	9484	9506	9528	2	4	7	9	11	13	15	17	20
.98	9550	9572	9594	9616	9638	9661	9683	9705	9727	9750	2	4	7	9	11	13	16	18	20
.99	9772	9795	9817	9840	9863	9886	9908	9931	9954	9977	2	5	7	9	11	14	16	18	20

	0	1	2	3	4	5	6	7	8	9	1	2	3	4	5	6	7	8	9
.00	1000	1002	1005	1007	1009	1012	1014	1016	1019	1021	0	0	1	1	1	1	2	2	2
.01	1023	1026	1028	1030	1033	1035	1038	1040	1042	1045	0	0	1	1	1	1	2	2	2
.02	1047	1050	1052	1054	1057	1059	1062	1064	1067	1069	0	0	1	1	1	1	2	2	2
.03	1072	1074	1076	1079	1081	1084	1086	1089	1091	1094	0	0	1	1	1	1	2	2	2
.04	1096	1099	1102	1104	1107	1109	1112	1114	1117	1119	0	1	1	1	1	2	2	2	2
.05	1122	1125	1127	1130	1132	1135	1138	1140	1143	1146	0	1	1	1	1	2	2	2	2
.06	1148	1151	1153	1156	1159	1161	1164	1167	1169	1172	0	1	1	1	1	2	2	2	2
.07	1175	1178	1180	1183	1186	1189	1191	1194	1197	1199	0	1	1	1	1	2	2	2	3
.08	1202	1205	1208	1211	1213	1216	1219	1222	1225	1227	0	1	1	1	1	2	2	2	3
.09	1230	1233	1236	1239	1242	1245	1247	1250	1253	1256	0	1	1	1	1	2	2	2	3
.10	1259	1262	1265	1268	1271	1274	1276	1279	1282	1285	0	1	1	1	1	2	2	2	3
.11	1288	1291	1294	1297	1300	1303	1306	1309	1312	1315	0	1	1	1	2	2	2	2	3
.12	1318	1321	1324	1327	1330	1334	1337	1340	1343	1346	0	1	1	1	2	2	2	3	3
.13	1349	1352	1355	1358	1361	1365	1368	1371	1374	1377	0	1	1	1	2	2	2	3	3
.14	1380	1384	1387	1390	1393	1396	1400	1403	1406	1409	0	1	1	1	2	2	2	3	3
.15	1413	1416	1419	1422	1426	1429	1432	1435	1439	1442	0	1	1	1	2	2	2	3	3
.16	1445	1449	1452	1455	1459	1462	1466	1469	1472	1476	0	1	1	1	2	2	2	3	3
.17	1479	1483	1486	1489	1493	1496	1500	1503	1507	1510	0	1	1	1	2	2	2	3	3
.18	1514	1517	1521	1524	1528	1531	1535	1538	1542	1545	0	1	1	1	2	2	2	3	3
.19	1549	1552	1556	1560	1563	1567	1570	1574	1578	1581	0	1	1	1	2	2	3	3	3
.20	1585	1589	1592	1596	1600	1603	1607	1611	1614	1618	0	1	1	1	2	2	3	3	3
.21	1622	1626	1629	1633	1637	1641	1644	1648	1652	1656	0	1	1	2	2	2	3	3	3
.22	1660	1663	1667	1671	1675	1679	1683	1687	1690	1694	0	1	1	2	2	2	3	3	3
.23	1698	1702	1706	1710	1714	1718	1722	1726	1730	1734	0	1	1	2	2	2	3	3	4
.24	1738	1742	1746	1750	1754	1758	1762	1766	1770	1774	0	1	1	2	2	2	3	3	4
.25	1778	1782	1786	1791	1795	1799	1803	1807	1811	1816	0	1	1	2	2	2	3	3	4
.26	1820	1824	1828	1832	1837	1841	1845	1849	1854	1858	0	1	1	2	2	3	3	3	4
.27	1862	1866	1871	1875	1879	1884	1888	1892	1897	1901	0	1	1	2	2	3	3	4	4
.28	1905	1910	1914	1919	1923	1928	1932	1936	1941	1945	0	1	1	2	2	3	3	4	4
.29	1950	1954	1959	1963	1968	1972	1977	1982	1986	1991	0	1	1	2	2	3	3	4	4
.30	1995	2000	2004	2009	2014	2018	2023	2028	2032	2037	0	1	1	2	2	3	3	4	4
.31	2042	2046	2051	2056	2061	2065	2070	2075	2080	2084	0	1	1	2	2	3	3	4	4
.32	2089	2094	2099	2104	2109	2113	2118	2123	2128	2133	0	1	1	2	2	3	3	4	4
.33	2138	2143	2148	2153	2158	2163	2168	2173	2178	2183	0	1	1	2	2	3	3	4	4
.34	2188	2193	2198	2203	2208	2213	2218	2223	2228	2234	1	1	2	2	3	3	4	4	5
.35	2239	2244	2249	2254	2259	2265	2270	2275	2280	2286	1	1	2	2	3	3	4	4	5
.36	2291	2296	2301	2307	2312	2317	2323	2328	2333	2339	1	1	2	2	3	3	4	4	5
.37	2344	2350	2355	2360	2366	2371	2377	2382	2388	2393	1	1	2	2	3	3	4	4	5
.38	2399	2404	2410	2415	2421	2427	2432	2438	2443	2449	1	1	2	2	3	3	4	4	5
.39	2455	2460	2466	2472	2477	2483	2489	2495	2500	2506	1	1	2	2	3	3	4	5	5
.40	2512	2518	2523	2529	2535	2541	2547	2553	2559	2564	1	1	2	2	3	4	4	5	5
.41	2570	2576	2582	2588	2594	2600	2606	2612	2618	2624	1	1	2	2	3	4	4	5	5
.42	2630	2636	2642	2649	2655	2661	2667	2673	2679	2685	1	1	2	2	3	4	4	5	6
.43	2692	2698	2704	2710	2716	2723	2729	2735	2742	2748	1	1	2	3	3	4	4	5	6
.44	2754	2761	2767	2773	2780	2786	2793	2799	2805	2812	1	1	2	3	3	4	4	5	6
.45	2818	2825	2831	2838	2844	2851	2858	2864	2871	2877	1	1	2	3	3	4	5	5	6
.46	2884	2891	2897	2904	2911	2917	2924	2931	2938	2944	1	1	2	3	3	4	5	5	6
.47	2951	2958	2965	2972	2979	2985	2992	2999	3006	3013	1	1	2	3	3	4	5	5	6
.48	3020	3027	3034	3041	3048	3055	3062	3069	3076	3083	1	1	2	3	4	4	5	6	6
.49	3090	3097	3105	3112	3119	3126	3133	3141	3148	3155	1	1	2	3	4	4	5	6	6

Index

307